Applied Mathematical Sciences | Volume 9

Applied Mathematical Sciences

F. John
PARTIAL DIFFERENTIAL EQUATIONS
ISBN 0-387-90021-7

L. Sirovich
TECHNIQUES OF ASYMPTOTIC ANALYSIS
ISBN 0-387-90022-5

J. Hale
FUNCTIONAL DIFFERENTIAL EQUATIONS
ISBN 0-387-90023-3

J. K. Percus
COMBINATORIAL METHODS
ISBN 0-387-90027-6

R. von Mises and K. O. Friedrichs
FLUID DYNAMICS
ISBN 0-387-90028-4

W. Freiberger and U. Grenander
A SHORT COURSE IN COMPUTATIONAL
PROBABILITY AND STATISTICS
ISBN 0-387-90029-2

A. C. Pipkin
LECTURES ON VISCOELASTICITY THEORY
ISBN 0-387-90030-6

G. E. O. Giacaglia
PERTURBATION METHODS IN NON-LINEAR SYSTEMS
ISBN 0-387-90054-3

K. O. Friedrichs

Applied Mathematical Sciences 9

Spectral Theory of Operators in Hilbert Space

Springer-Verlag
New York · Heidelberg · Berlin

Applied Mathematical Sciences

EDITORIAL STATEMENT

The mathematization of all sciences, the fading of traditional scientific boundaries, the impact of computer technology, the growing importance of mathematical-computer modelling and the necessity of scientific planning all create the need both in education and research for books that are introductory to and abreast of these developments.

The purpose of this series is to provide such books, suitable for the user of mathematics, the mathematician interested in applications, and the student scientist. In particular, this series will provide an outlet for material less formally presented and more anticipatory of needs than finished texts or monographs, yet of immediate interest because of the novelty of its treatment of an application or of mathematics being applied or lying close to applications.

The aim of the series is, through rapid publication in an attractive but inexpensive format, to make material of current interest widely accessible. This implies the absence of excessive generality and abstraction, and unrealistic idealization, but with quality of exposition as a goal.

Many of the books will originate out of and will stimulate the development of new undergraduate and graduate courses in the applications of mathematics. Some of the books will present introductions to new areas of research, new applications and act as signposts for new directions in the mathematical sciences. This series will often serve as an intermediate stage of the publication of material which, through exposure here, will be further developed and refined and appear later in one of Springer-Verlag's other mathematical series.

MANUSCRIPTS

The Editors welcome all inquiries regarding the submission of manuscripts for the series. Final preparation of all manuscripts will take place in the editorial offices of the series in the Division of Applied Mathematics, Brown University, Providence, Rhode Island.

SPRINGER-VERLAG NEW YORK INC., 175 Fifth Avenue, New York, N.Y. 10010

Printed in U.S.A.

K. O. Friedrichs

Spectral Theory of Operators in Hilbert Space

Springer-Verlag New York · Heidelberg · Berlin
1973

K. O. Friedrichs
New York University
Courant Institute of Mathematical Sciences

AMS Classification
47A05, 47A10, 47B25, 47B40

Library of Congress Cataloging in Publication Data

Friedrichs, Kurt Otto.
Spectral theory of operators in Hilbert space.

(Applied mathematical sciences, v. 9)
1. Hilbert space. 2. Spectral theory (Mathematics)
3. Operator theory. I. Title. II. Series.
QA1.A647 vol. 9 [QA322.4] 510'.8s [515'.73] 73-13721

Printed in the United States of America.

ISBN 0-387-90076-4 Springer-Verlag New York Heidelberg Berlin
ISBN 3-540-90076-4 Springer-Verlag Berlin Heidelberg New York

PREFACE

The present lectures intend to provide an introduction to the spectral analysis of self-adjoint operators within the framework of Hilbert space theory. The guiding notion in this approach is that of spectral representation. At the same time the notion of function of an operator is emphasized.

The formal aspects of these concepts are explained in the first two chapters. Only then is the notion of Hilbert space introduced. The following three chapters concern bounded, completely continuous, and non-bounded operators. Next, simple differential operators are treated as operators in Hilbert space, and the final chapter deals with the perturbation of discrete and continuous spectra.

The preparation of the original version of these lecture notes was greatly helped by the assistance of P. Rejto. Various valuable suggestions made by him and by R. Lewis have been incorporated.

The present version of the notes contains extensive modifications, in particular in the chapters on bounded and unbounded operators.

February, 1973 K.O.F.

TABLE OF CONTENTS

Spectral Theory of Operators in Hilbert Space

CHAPTER I

SPECTRAL REPRESENTATION

1. Three Typical Problems

The problem of the spectral representation of a linear operator arises as a natural generalization of the problem of the transformation of a quadratic form to principal axes. In this section we shall discuss this and two well-known analogous problems in a preliminary fashion.

Example 1. Suppose a quadratic form in n real variables $\xi_1,\ldots,\xi_n$ is given as the expression

$$\sum_{\sigma,\sigma'=1}^{n} a_{\sigma\sigma'}\xi_\sigma\xi_{\sigma'} ,$$

in which the coefficients $a_{\sigma\sigma'} = a_{\sigma'\sigma}$ are real numbers. The n numbers ξ_σ may be regarded as the components of a vector $\Xi = (\xi_\sigma)$ with respect to a coordinate system in an n-dimensional Euclidean space. Then the problem is to rotate this coordinate system so that the quadratic form assumes the simple form

$$(1.1) \qquad Q(\Xi) = \sum_{\sigma,\sigma'=1}^{n} a_{\sigma\sigma'}\xi_\sigma\xi_{\sigma'} = \sum_{\mu=1}^{n} \alpha_\mu\eta_\mu^2 .$$

Here $\eta_1,\ldots,\eta_n$ are the components of the vector Ξ with respect to the new system, connected with the coordinates ξ_σ through a transformation given by linear relations

$$(1.2) \qquad \xi_\sigma = \sum_{\mu=1}^{n} u_{\sigma\mu}\eta_\mu$$

$$(1.2)^* \qquad \eta_\mu = \sum_{\sigma=1}^{n} v_{\mu\sigma}\xi_\sigma \; .$$

The requirement that the new coordinate system be obtained through rotation from the original system is expressed by the condition that the square of the magnitude of the vector is the sum of the squares of the coordinates with respect to each coordinate system:

$$(1.3) \qquad \sum_{\sigma=1}^{n} \xi_\sigma^2 = \sum_{\mu=1}^{n} \eta_\mu^2 \; .$$

The numbers $\alpha_1,\ldots,\alpha_n$ entering identity (1.1) are called "eigen-values" of the quadratic form $Q\{\Xi\}$ since this form assumes these values for the unit vectors of the new coordinate system. These are the vectors for which all components $\eta_1,\ldots,\eta_n$ equal 0 except one which equals 1. Specifically, we denote by $H^{(\mu')}$ the vector with the new coordinates

$$\eta_\mu = \delta_{\mu\mu'} \; .$$

Here we have employed the "Kronecker symbol"

$$\delta_{\mu\mu'} = 0 \quad \text{for} \quad \mu \neq \mu'. \quad \delta_{\mu\mu} = 1.$$

The ξ coordinates of the vector $H^{(\mu)}$ are $\xi_\sigma = u_{\sigma\mu}$ as seen from (1.2),

$$(1.4) \qquad H^{(\mu)} = (u_{\mu\sigma}) .$$

The vectors $H^{(\mu)}$ are also called unit "eigen-vectors" of the quadratic form; any multiple $cH^{(\mu)} \neq 0$ of such a vector will be called "eigen-vector".

Before indicating how one could establish a transformation

(1.2), (1.2)* such that identities (1.1), (1.3) hold we shall assume that there is such a transformation and derive various consequences from this assumption.

If the relation (1.3) holds for all vectors Ξ it also holds for the linear combination $c\,\Xi + c_1\,\Xi^{(1)}$ of any two vectors Ξ, $\Xi^{(1)}$ with arbitrary coefficients c, c_1. Identifying the mixed terms in the relation

$$\sum_{\sigma} (c\xi_{\sigma} + c_1\xi_{\sigma}^{(1)})^2 = \sum_{\sigma} (c\eta_{\mu} + c_1\eta_{\mu}^{(1)})^2,$$

in which η_{σ} and $\eta_{\sigma}^{(1)}$ are the new coordinates of Ξ and $\Xi^{(1)}$, we obtain the relation

$$(1.3)' \qquad \sum_{\sigma} \xi_{\sigma}\xi_{\sigma}^{(1)} = \sum_{\mu} \eta_{\mu}\eta_{\mu}^{(1)} ,$$

which is thus seen to hold for all vectors Ξ, $\Xi^{(1)}$. Here and in the following $\sum_{\sigma}, \sum_{\mu}$ stands for $\sum_{\sigma=1}^{n}, \sum_{\mu=1}^{n}$.

In a similar manner one derives from formula (1.1) the identity

$$(1.1)' \qquad \sum_{\sigma,\sigma'} \xi_{\sigma}a_{\sigma\sigma'}\xi_{\sigma'}^{(1)} = \sum_{\mu} \alpha_{\mu}\eta_{\mu}\eta_{\mu}^{(1)} .$$

Thus the identities (1.1), (1.3) concerning quadratic forms imply corresponding identities concerning the corresponding "bilinear forms".

We can draw further conclusions from relations (1.3)', (1.1)'. To this end let us take the eigenvector $H^{(\mu')}$ for $\Xi^{(1)}$ in these relations. Since, by (1.4), the ξ-coordinates of $H^{(\mu)}$ are $u_{\sigma\mu}$ and the η-coordinates of $H^{(\mu')}$ are $\delta_{\mu\mu'}$, we find

$$(1.3)'' \qquad \sum_{\sigma} \xi_{\sigma}u_{\sigma\mu} = \eta_{\mu} ,$$

(1.1)"
$$\sum_{\sigma\sigma'} \xi_\sigma a_{\sigma\sigma'} u_{\sigma'\mu} = \alpha_\mu \eta_\mu,$$

having replaced μ' by μ. Expressing η_μ in terms of the ξ by (1.2)* and identifying the coefficients of ξ_σ on both sides we obtain the identities

(1.3)'''
$$u_{\sigma\mu} = v_{\mu\sigma}$$

(1.1)'''
$$\sum_{\sigma'} a_{\sigma\sigma'} u_{\sigma'\mu} = \alpha_\mu v_{\mu\sigma}.$$

Relation (1.3)" allows one to determine the transformation matrix $(u_{\sigma\mu})$ once its inverse $(v_{\mu\sigma})$ is known and vice versa. Inserting this relation into (1.1)" we obtain the important formula

(1.5)
$$\sum_{\sigma'} \alpha_{\sigma\sigma'} u_{\sigma'\mu} = a_\mu u_{\sigma\mu},$$

which we shall interpret presently.

In an extensive investigation of transformations to principal axes one must shift the emphasis from the quadratic form to the <u>operator</u> associated with it. This operator, A, transforms the vector Ξ with the components $\xi_1, \ldots, \xi_n$ into the vector, $A\,\Xi$, with the components $\sum_{\sigma'} a_{\sigma\sigma'} \xi_{\sigma'}$. Thus

(1.6)
$$A\,\Xi = \left(\sum_{\sigma'} a_{\sigma\sigma'} \xi_{\sigma'}\right).$$

What is the effect of this operator in terms of the new coordinates η_μ? To find this out we express $\xi_{\sigma'}$ in terms of the η_μ by (1.2) and determine the η_μ-coordinate of $\left(\sum_{\sigma'} a_{\sigma\sigma'} \xi_{\sigma'}\right)$ by (1.2)*. The result is

$$\sum_{\mu}\left[\sum_{\sigma,\sigma'} v_{\mu\sigma}a_{\sigma\sigma'}u_{\sigma'\mu'}\right]\eta_{\mu'} \ .$$

By virtue of the relation (1.5), the expression in the bracket equals

$$\sum_{\sigma} v_{\mu\sigma}u_{\sigma\mu'}\alpha_{\mu}\delta_{\mu\mu'}$$

since (1.2) and (1.2)* imply $\sum_{\sigma} v_{\mu\sigma}u_{\sigma\mu'} = \delta_{\mu\mu'}$. Hence the η_{μ}-coordinate of $A\,\underline{\overline{\ -\ }}$ is simply $\alpha_{\mu}\eta_{\mu}$.

Thus we have found that <u>in the new coordinate system</u>, in which the vector $\underline{\overline{\ -\ }}$ has the component η_{μ}, <u>the vector</u> $A\,\underline{\overline{\ -\ }}$ <u>has the components</u> $\alpha_{\mu}\eta_{\mu}$.

This fact leads to a different formulation of the problem of transformation to principal axes. Instead of requiring that the quadratic form Q should become a sum or difference of squares, as given by (1.1), we may require that in terms of the new coordinates the given operator A should simply consist in multiplying each coordinate by a number, called an "eigen-value" of this operator. This formulation will lead directly to the notion of "spectral representation".

To explain the significance of the property of the operator A just derived let us apply this operator on the vector $H^{(\mu)}$, whose η-components are all zero except $\eta_{\mu} \neq 0$. Our property evidently implies that the η-components of $AH^{(\mu)}$ are all zero except the μth component, which is $\alpha_{\mu}\eta_{\mu}$. An obvious consequence of this fact is the relation

$$AH^{(\mu)} = \alpha_{\mu}H^{(\mu)} . \qquad (1.5)'$$

Thus, when applied to the vector $H^{(\mu)}$, the operator A acts like a multiplier with the value α_{μ}. It is for this reason that the number α_{μ} and the vector $H^{(\mu)}$ are called an "eigen-value" and "eigen-vector" of the operator A.

Note, in view of (1.4), that equation (1.5) is nothing but the expression of equation (1.5)' in terms of ξ-coordinates.

So far we have derived a number of properties from the assumption that transformation of the form Q, or the operator A, is possible. What about the problem of proving that there is such a transformation? One possible approach to doing this starts with equation (1.5). Writing ξ_σ, in place of $u_{\sigma\mu}$ and α in place of α_μ, this equation takes the form

$$\sum_{\sigma'} a_{\sigma\sigma'}\xi_{\sigma'} = \alpha\xi_\sigma \; , \qquad (1.5)''$$

which shows that all vectors $\Xi = H^{(\mu)}$ and eigen-value $\alpha = \alpha_\mu$ satisfy the same equation. Once α is chosen this equation may be regarded as a homogeneous system of equations for the n unknowns $\xi_1,\ldots,\xi_n$. The condition that this system have a solution other than $\xi_1,\ldots,\xi_n = 0$ is that its determinant vanishes:

$$\det(a_{\sigma\sigma'} - \alpha\delta_{\sigma\sigma'}) = 0. \qquad (1.7)$$

This condition may be regarded as an equation of the nth degree for α; it can be used to determine the eigen-values $\alpha = \alpha_\mu$. Having found this, vectors $\Xi = H^{(\mu)}$ can be found whose components satisfy equation (1.5)". Moreover, it is possible to find n such eigen-vectors which have all the properties discussed and whose ξ-component $u_{\sigma\mu}$ are the coefficients of the derived transformation to principal axes.

We shall not follow up this approach since it is not suitable for extension to problems involving a space of infinitely many dimensions.

Problem 2. In our second problem we begin with the transformation to principal axes; only later on we shall interpret the vectors having the direction of principal axes as eigen-vectors of an associated operator.

We consider complex-valued functions $\phi(s)$ of the real variable s which are periodic with the period 2π, so that

$$\phi(\pi) = \phi(-\pi), \quad \frac{d}{ds}\phi(\pi) = \frac{d}{ds}\phi(-\pi). \tag{1.8}$$

Actually, therefore, the functions $\phi(s)$ need be defined only for $-\pi \leq s \leq \pi$. Of these functions we assume at present that they have continuous derivatives up to the second order. As is well-known, such functions $\phi(s)$ admit a uniformly convergent Fourier expansion

$$\phi(s) = \sum_{\mu} e^{is\mu}\eta_{\mu} \tag{1.9}$$

in which the summation runs over all integers μ,

$$-\infty < \mu < \infty, \quad \mu \text{ integer}.$$

The Fourier coefficients η_{μ} are given by the formula

$$\eta_{\mu} = \frac{1}{2\pi}\int_{-\pi}^{\pi} e^{-i\mu s}\phi(s)\,ds. \tag{1.9*}$$

The analogy of these two formulas with the formula (1.2) and (1.2)* is apparent. There is even an analogue with formula (1.10), viz,

$$\frac{1}{2\pi}\int_{-\pi}^{\pi} |\phi(s)|^2 ds = \sum_{\infty<\mu<\infty} |\eta_{\mu}|^2. \tag{1.10}$$

We may try to push the analogy further and consider the values of the function ϕ for $-\pi < s \leq \pi$ as well as the values of η for integral μ as the coordinates, with respect to different coordinate systems, of a single entity, a "vector" Φ. Formula (1.10) then expresses the square of the "magnitude" of this vector. In doing so we have no trouble in introducing unit vectors $H^{(\mu)}$ for the new coordinate system as vectors for which $\eta_\mu = 1$ and $\eta_{\mu'} = 0$ for $\mu' \neq \mu$. The functions $\phi(s)$ associated with these vectors are the "eigen-functions"

$$\phi^{(\mu)}(s) = e^{i\mu s}, \tag{1.11}$$

as seen from (1.9). We do have some trouble, however, in trying to introduce unit vectors with respect to an original coordinate system. What to do about this difficulty will also be discussed in Section 2. At present we simply say that the function $\phi(s)$ is the representative of the vector Φ with respect to one coordinate system, while the sequence η_μ is its representative with respect to another coordinate system, never mind that these coordinate systems are not defined in terms of unit-vectors.

In another respect we have no trouble in pushing the analogy further. It is easy to find an operator whose eigen-values are just the numbers μ and whose eigen-vectors are just the vectors $H^{(\mu)}$. This is the differential operator

$$-i \frac{d}{ds} \tag{1.12}$$

acting on the functions $\phi(s)$. Regarding the function $\phi(s)$ as representative of a vector Φ we regard the operator $-id/ds$ as representative of an operator - denoted by M - acting on such vectors.

By virtue of the strong differentiability assumptions we have made on the functions Φ we may differentiate relation (1.9) under the summation sign and carry out integration by parts on (1.9)*. Thus we obtain the relations

$$-i \frac{d}{ds} \phi(s) = \sum_{\mu} e^{is\mu} \mu\eta_{\mu}, \tag{1.13}$$

$$\mu\eta_{\mu} = \frac{1}{2\pi} \int_{-\pi}^{\pi} e^{-is\mu}(-i \frac{d}{ds} \phi(s))ds. \tag{1.13'}$$

They show that, in terms of η-components, application of the operator M just means multiplying each component η_{μ} by the number μ. In particular, if the vector Φ is specialized to be the vector $H^{(\mu)}$ whose η-components all vanish except η_{μ}, we have

$$MH^{(\mu)} = \mu H^{(\mu)} . \tag{1.14}$$

Therefore, the number μ is an eigen-value of the operator M and the vector $H^{(\mu)}$ is the associated eigen-vector.

In terms of the eigen-functions $\phi^{\mu}(s) = e^{is\mu}$, which correspond to the eigen-vectors $H^{(\mu)}$, relation (1.14) assumes the form

$$-i \frac{d}{ds} \phi^{(\mu)}(s) = \mu\phi^{(\mu)}(s). \tag{1.14'}$$

One of the major differences between the situation in the present and the first example is the fact that now the spectrum consists of infinitely many eigen-values, instead of a finite number, as in the first example.

Problem 3. In the third problem, which we proceed to discuss now, the notion of spectrum as such is affected. This example is concerned with the Fourier integral in contrast with the Fourier series.

Again we consider complex-valued functions $\phi(s)$ of a real variable s which now is to range over the whole s-axis, $-\infty < s < \infty$.

We again assume, for convenience, that the functions $\phi(s)$ have continuous derivatives up to the second order. Also, we require that the functions tend to zero as $|s| \to \infty$ sufficiently rapidly. We adopt the customary condition

$$\int_{-\infty}^{\infty} |\phi(s)|ds < \infty \text{ together with } \phi(s) \to 0 \text{ as } |s| \to \infty \tag{1.15}$$

supplementing it by the condition

$$\int_{-\infty}^{\infty} \left|\frac{d}{ds}\,\phi(s)\right|ds < \infty \text{ together with } \frac{d\phi}{ds}(s) \to 0 \text{ as } |s| \to \infty. \tag{1.15'}$$

Actually, for our present purposes it does not matter much in which way the function $\phi(s)$ is required to die out "sufficiently rapidly."

Functions $\phi(s)$ which behave in the manner described admit a Fourier integral representation

$$\phi(s) = \int_{-\infty}^{\infty} e^{is\mu}\eta(\mu)\,d\mu \tag{1.16}$$

in terms of a continuous "transform" $\eta(\mu)$, given by

$$\eta(\mu) = \frac{1}{2\pi}\int_{-\infty}^{\infty} e^{-i\mu s}\phi(s)\,ds. \tag{1.16*}$$

The functions $\phi(s)$ and $\eta(\mu)$ are also connected through the relation

$$\frac{1}{2\pi}\int_{-\infty}^{\infty} |\phi(s)|^2 ds = \int_{-\infty}^{\infty} |\eta(\mu)|^2 d\mu. \tag{1.17}$$

Again we may try to interpret the values of the function $\phi(s)$ and those of $\eta(\mu)$ as components of a "vector" Φ with respect to an old and a new coordinate system. But, this time we cannot introduce unit coordinate vectors with respect to either coordinate system. There is no function $\phi(s)$ among those considered, whose transform $\eta(\mu)$ is different from zero for only one value of μ. In particular, the function $\phi^{(\mu)}(s) = e^{is\mu}$ is not one of these functions; its transform is not defined in a proper sense. If the transform $\eta_{\mu'}(\mu)$ of $e^{is\mu'}$ were defined it would vanish for $\mu \neq \mu'$ and be infinite for $\mu = \mu'$.

In spite of this awkwardness it would be possible to enlarge the class of functions so as to include those that may serve as coordinate unit vectors in the present problem. We shall eventually discuss this extension but not do so in the development of the basic theory; for, either the simplicity or the completeness of the general theory would be destroyed. We shall employ a generalization of the notion of function (in Chapter III); but a rather restricted one. We shall be guided by the requirement that the "space integrals" entering formula (1.17) remain finite for the generalized functions.

In the first two chapters we shall confine ourselves to describing the notion of transformation to principal axes in terms which do not require an extension of the notion of function.

Returning to our third problem we maintain that the differential operator

$$- i \frac{d}{ds} ,$$

plays essentially the same role as in the second example, except that this operator now applies to functions defined all over the s-axis. Let $\eta(\mu)$ be the transform of the function $\phi(s)$. Then, we maintain the transform of the function $-i \frac{d}{ds} \phi(s)$ is the function $\mu\eta(\mu)$. This is immediately verified by differentiation under the integral sign in (1.16) or integration by parts in (1.16)*.

Thus, if we regard the function $\eta(\mu)$ as the representative of a function $\phi(s)$, we find that the function $\mu\eta(\mu)$ is the representative of function $-i\frac{d}{ds}\phi(s)$. This description will lead to the notion of "spectral representation".

Although we no longer can speak of eigen-values and eigen-vectors or -functions in the proper sense, we recognize that the values μ play a role analogous to that of an eigen-value. The set of all these values μ is again called the spectrum of the operator M, but this spectrum is said to be continuous in contrast to the spectra of the first two examples, which are called discrete. The "improper eigen-functions" $e^{is\mu}$ will not be regarded as functions on which the operator may act; these functions will be relegated to the role they play in the transformation formulas (1.16) and (1.16)*.

In spite of these differences between the cases with discrete and continuous spectrum the notion of spectral representation can be formulated in general fashion so as to include all cases. In order to do this we first recall the notions of "linear space" and "linear operator".

2. Linear Space and Functional Representation. Linear Operators.

In this section we shall give a brief resume of the basic facts of linear spaces as much as we need them for our purposes.

A linear space $\mathfrak{S}$ consists of elements, called "vectors", Φ for which linear combinations are defined. That means, to every pair $\Phi, \Phi^{(1)}$ of vectors and every pair c, c_1 of numbers a vector denoted by

$$c\Phi + c_1\Phi^{(1)}$$

is assigned in such a way that the customary rules of algebraic operations hold; that is, the operation "addition" should be commutative and associative, and the operation "multiplication" by a number should be distributive and associative. There should be a "zero" vector 0, such that

$$\Phi + 0 = \Phi, \quad 0\Phi = 0, \quad c0 = 0.$$

At present we leave open whether the coefficients c may be complex or should be real; i.e., at present we consider complex and real linear spaces together. Eventually we shall confine ourselves to complex spaces. When we speak of a "space" in the following we shall always mean a "linear space".

We say a space $\mathfrak{S}$ has finite <u>dimension</u>* if it contains a finite number, n say, of vectors $\Xi^{(1)}, \dots, \Xi^{(n)}$ such that every vector Φ in $\mathfrak{S}$ can be written as a linear combination of them:

$$\Phi = \xi_1 \Xi^{(1)} + \dots + \xi_n \Xi^{(n)} . \tag{2.1}$$

We then say, the n vectors Ξ span the space $\mathfrak{S}$. The dimension of the space $\mathfrak{S}$ is exactly n if it cannot be spanned by less than n vectors. We then say the n vectors Ξ generate a coordinate system. The numbers $\xi_1, \dots, \xi_n$ are called the components of Φ with respect to this system. If $\mathfrak{S}$ is real the components ξ are real, otherwise they may be complex.

A space which cannot be spanned by a finite number of vectors has "infinite dimension". The manifold of vectors Φ considered in our second example is such a space of infinite dimension (for proof see a later section). The number of components η_μ with respect to the "new" coordinate system is infinite, specifically, a denumerable set of numbers. The set of "components" $\phi(s)$ with respect to the original description is non-denumerable, since the set of all numbers

---

*We prefer simply to say "finite dimension" instead of "a finite number of dimensions", although the latter expression corresponds with the original meaning of the term dimension.

s between $-\pi$ and π is non-denumerable. For this reason we do <u>not</u> simply say the dimension of the space is the cardinal number of the set of components.

Let $\mathfrak{S}$ be any space and let $[\xi]$ be a space consisting of functions $\xi(s)$, a "function space" for short. We then say <u>the vectors of the space</u> $\mathfrak{S}$ <u>are represented by the functions of the</u> space $[\xi]$ if there is a one-to-one correspondence of the vectors Φ of $\mathfrak{S}$ and the functions $\xi(s)$ of $[\xi]$ in such a way that the linear combination $c\Phi + c_1\Phi^{(1)}$ is represented by the function $c\xi(s) + c_1\xi^{(1)}(s)$ provided Φ and $\Phi^{(1)}$ are represented by $\xi(s)$ and $\xi^{(1)}(s)$. The functions $\xi(s)$ will be called the "representers" of the vectors Φ. Note that the required one-to-one character of the correspondence guarantees that $\xi(s) \equiv 0$ implies $\Phi = 0$. A representation of the Φ by functions $\xi(s)$ will be referred to as a "functional representation"; it will symbolically be expressed by the formula

$$\Phi \underset{\xi}{\longleftrightarrow} \{\xi(s)\}.$$

The double arrow is to indicate the one-to-one character of the correspondence. The subscript is to indicate the specific space of representing functions involved.

Evidently, the vectors Φ of a space $\mathfrak{S}$ admit different functional representations. For instance, the vectors Φ of the third example are represented by functions $\eta(\mu)$; thus we may write

$$\Phi \underset{\eta}{\longleftrightarrow} \{\eta(\mu)\}.$$

Originally, these vectors were given as functions $\phi(s)$; giving vectors as functions automatically establishes a functional representation. Instead of writing $\Phi \underset{\phi}{\longleftrightarrow} \phi(s)$, we prefer to write simply

$$\Phi = \phi$$

in such a case.

The vectors of the second example were also given as functions, $\Phi = \emptyset$; with respect to the new coordinate system they were represented by infinitely many numbers, η_μ, $\mu = 0, \pm 1, \pm 2, \ldots$. Here η_μ may be regarded as a function of the variable μ, never mind that this variable was restricted to the integers. Thus we may write

$$\Phi \underset{\eta}{\longleftrightarrow} \{\eta_\mu\}.$$

In our first example the vectors Φ were given with the aid of their components $\xi_1, \ldots, \xi_n$ with respect to the original coordinate system. We may just as well say they were given through a particular representation and write accordingly $\Phi \underset{\xi}{\longleftrightarrow} \{\xi_\sigma\}$ or simply $\Phi = \{\xi_\sigma\}$ Note that the set of components of a vector Φ then plays the role of a representer, a function of the variable σ which runs over the numbers $1, 2, \ldots, n$.

Next we discuss the notion of <u>operator</u>. A linear operator is a transformation, or mapping, which transforms vectors Φ in the space $\mathfrak{S}$ into vectors $A\Phi$ in $\mathfrak{S}$ in such a way that the relation

$$A(c\Phi + c_1\Phi^{(1)}) = cA\Phi + c_1 A\Phi^{(1)} \tag{2.2}$$

holds for all c, c_1 and all Φ, Φ^1. We shall always omit the qualification "linear"; that is to say whenever we say "operator" we mean "linear operator".

Such an operator need not be defined in the whole space $\mathfrak{S}$. For example, if $\mathfrak{S}$ is the space of continuous functions $\mathfrak{S}(s)$ the operator d/ds is defined only in the subspace of functions with a continuous derivative. The manifold of vectors Φ in $\mathfrak{S}$ for which an operator A is defined is called its "domain" $\mathfrak{S}_A$; by definition $\mathfrak{S}_A$ is a linear space. Instead of saying that a vector Φ belongs to the space $\mathfrak{S}_A$, we shall frequently simply say that Φ "admits" A, or that A is applicable to Φ.

An operator A defined in a space $\mathfrak{S}$ of finite dimension, n, is represented by a matrix, when the vectors Φ are represented by components, $\Phi \underset{\xi}{\longleftrightarrow} \{\xi_\sigma\}$. For, let $\Phi^{(\sigma')}$ be the vector with the components $\xi_{\sigma'} = 1$, $\xi_\sigma = 0$ for $\sigma \neq \sigma'$, and denote by $a_{1\sigma}, \ldots, a_{n\sigma}$, the n components of the vector $A\Phi^{(\sigma')}$. Then, by (2.1) and (2.2) we have

$$\Phi = \sum_\sigma \xi_\sigma \Phi^{(\sigma)}$$

and hence

$$A\Phi = \sum_\sigma \left(\sum_{\sigma'} \xi_{\sigma'} a_{\sigma\sigma'}\right) \Phi^{(\sigma)} .$$

Thus the representation $\Phi = \{\xi_\sigma\}$ implies the representation

$$A\Phi = \left(\sum_{\sigma'} a_{\sigma\sigma'} \xi_{\sigma'}\right) .$$

Clearly, the operator A is represented by the matrix $(a_{\sigma\sigma'})$.

When the vector Φ is represented by functions $\phi(s)$ of a continuous variable, the operator A will be represented by a "functional operator" which may be an integral operator or a differential operator such as i d/ds or it may be of a more involved nature. Later on we shall study such functional operators in detail.

3. Spectral Representation.

We are now in a position to say what is meant by a "spectral" representation of an operator A defined in a subspace $\mathfrak{S}_A$ of a space $\mathfrak{S}$.

First we explain the notion of simple spectral representation. It consists in the (linear) representation of the vectors Φ in the space $\mathfrak{S}$ by certain functions $\eta(\alpha)$ of a real* variable α in such a way that the vector $A\Phi$, for Φ in $\mathfrak{S}_A$, corresponds to the function $\alpha\eta(\alpha)$. In formulas:

$$(3.1) \qquad \Phi \underset{\eta}{\longleftrightarrow} \{\eta(\alpha)\} \qquad \text{for all } \Phi \text{ in } \mathfrak{S}$$

implies

$$(3.1)_A \qquad A\Phi \underset{\eta}{\longleftrightarrow} \{\alpha\eta(\alpha)\} \qquad \text{for all } \Phi \text{ in } \mathfrak{S}_A .$$

Thus the operator A is simply represented by multiplication by the independent variable of the functions $\eta(\alpha)$.

The variable α will be called the "spectral" variable; the domain of the functions $\eta(\alpha)$, i.e., the set S of values α for which the functions $\eta(\alpha)$ are defined will -- temporarily -- be called the "spectrum" of the operator A. Note that this domain may consist of the whole α-axis or it may be part of it. In fact, it may consist just of distinct points, as in our first and second examples. The notion of "spectrum" is introduced here with reference to a spectral representation. Later on, in Chapters V and VI we shall define this term differently, independently of such a representation.

Let us ask which role eigen-values and eigen-vectors play in connection with a spectral representation. Eigen-values are numbers α' to which there is a vector $H' \neq 0$ in $\mathfrak{S}_A$ such that $AH' = \alpha' H'$. For the representer $\eta'(\alpha)$ of H' this relation becomes

---

*In some later stage the restriction to real variables will be removed temporarily.

$$\alpha\eta'(\alpha) = \alpha'\eta'(\alpha),$$

which implies that

$$\eta'(\alpha) = 0 \quad \text{for} \quad \alpha \neq \alpha'.$$

Conversely, suppose that the domain $\mathfrak{S}_A$ of A contains a vector $H' \neq 0$ whose representer $\eta'(\alpha)$ vanishes except for a particular value α'. Then $\alpha\eta'(\alpha) = \alpha'\eta'(\alpha)$ for all values α in the spectrum and hence relation $AH' = \alpha'H'$ holds. In other words, H' is an eigen-vector with eigen-value α'. Clearly, α' belongs to the spectrum; for, else $\eta'(\alpha) = 0$ for all α in the spectrum which would mean $H' = 0$.

It is quite convenient to use the term "eigen-value" for any value in the spectrum. But then one must distinguish between the "proper" eigen-values, i.e., eigen-values in the strict sense as described above, and the others, "improper" ones. Proper eigen-values are also called point eigen-values. The set of improper eigen-values is also called the "continuous" part of the spectrum.

Naturally one may wonder whether or not there is an analogue of an eigen-vector for a continuous spectrum. Such an analogue can readily be introduced if it is assigned - not to a single, improper, eigen-value - but to an interval $\Delta\alpha$ on the α-axis. To every such interval we assign as "eigen-space" the space of all those vectors in $\mathfrak{S}_A$ whose representers $\eta(\alpha)$ vanish identically outside of the interval $\Delta\alpha$. The vectors in this eigen-space - except the zero-vector - will then be called eigen-vectors associated with the interval $\Delta\alpha$.

For example, in our third problem all functions $\phi(s)$ of the form

(3.2) $$\int_{\Delta\mu} e^{is\mu}\eta(\mu)\,d\mu$$

are eigen-functions of the operator $-\,id/ds$ associated with the interval $\Delta\mu$.

The set of eigen-vectors associated with intervals includes those associated with points provided one regards a single point as an interval. If the eigen-space associated with a point is empty (except for the zero-vector) this point is not an eigen-value; otherwise it is.

Incidentally, intervals $\Delta\alpha$ differing only in the inclusion of one or two end points must be counted as different in the present context; for, a non-empty eigen-space may belong to such an end point.

Still, one may wonder whether one could not introduce eigen-vectors associated strictly with single improper eigen-values. For example, in connection with our third example, it might be suggested to introduce the functions $e^{is\mu'}$ as "improper eigen-functions" since they satisfy the relation $-\,i(d/ds)e^{is\mu'} = \mu' e^{is\mu'}$. Of course, since these functions do not belong to $\mathfrak{S}_A$, the domain of the operator introduced, it would be necessary to enlarge this domain. Such an enlargement would have awkward consequences, as was described when this example was discussed. Therefore, we shall at present not employ such an extension.

Improper eigen-functions will, however, retain a role in the explicit description of the spectral transformation.

The notion of simple spectral representation which we have described is too narrow since it does not cover cases with a multiple spectrum. For this reason it is necessary to introduce more general notions of spectral representation. It is expedient to do so also in order to attain greater flexibility in handling concrete spectral problems. Appropriate generalizations may be carried out in two directions.

A first generalization arises if the vectors Φ of the space $\mathfrak{S}$ are represented by functions $\eta(\mu)$ of an "auxiliary" spectral variable μ in such a way that the vectors $A\Phi$, for Φ in $\mathfrak{S}_A$, are represented by the functions $\alpha(\mu)\eta(\mu)$; here $\alpha(\mu)$ is an appropriate function of μ. We express such a representation by the formulas

(3.3) $$\Phi \underset{\eta}{\longleftrightarrow} \{\eta(\mu)\}.$$

$(3.3)_A$ $$A\Phi \underset{\eta}{\longleftrightarrow} \{\alpha(\mu)\eta(\mu)\}.$$

We then speak of an "indirect" spectral representation, while the representation will be called "direct" if $\mu = \alpha$. The variable α will be restricted to an appropriate domain; the range of the function $\alpha(\mu)$, the "spectral function" is then the spectrum of the operator A.

The spectral representation implied by the principal axis transformation given in the first example is evidently an indirect one. Every vector $\overline{\underline{-}}$ is represented by a function $\eta(\mu) = \eta_\mu$ of an auxiliary spectral variable μ whose domain consists of the n points $\mu = 1,\ldots,n$.

As another example we consider the operator

(3.4) $$M^2 = -\frac{d^2}{ds^2}$$

applicable on functions $\emptyset(s)$ defined for $-\pi \leq s \leq \pi$ which possess continuous third derivatives and satisfy the periodicity conditions $\emptyset, d\emptyset/ds, d^2\emptyset/ds^2|_{-\pi}^{\pi} = 0$. Suppose we represent these functions $\emptyset(s)$ by their Fourier coefficients η_μ according to (1.9), (1.9)*; then the functions $M^2\Phi(s) = -d^2\Phi(s)/ds^2$ are represented by their Fourier coefficients $\mu^2\eta_\mu$; i.e.,

$$(3.5) \qquad \phi(s) \longleftrightarrow \{\eta_\mu\}$$

implies

$$(3.5)_{M^2} \qquad -\frac{d^2}{ds^2}\phi(s) \longleftrightarrow \{\mu^2\eta_\mu\}.$$

This representation is evidently an indirect spectral representation with the auxiliary variable μ and the spectral function $\alpha(\mu) = \mu^2$. It is seen that the spectrum of this operator consists of the eigen-values $0,1,4,9,\ldots$, those greater than zero occurring twice, i.e., having two linearly independent eigen-functions.

The notion of eigen-space associated with an interval $\Delta\alpha$, which was introduced in connection with the simple spectral representation, can immediately be carried over to the case of an indirect representation. We assign to $\Delta\alpha$ the set of $\Delta\mu$ of those values of μ for which the values of $\alpha(\mu)$ lie in $\Delta\alpha$. Then the eigen-space associated with the interval $\Delta\alpha$ consists of all those vectors Φ in $\mathfrak{S}$ whose representers $\eta(\mu)$ vanish when μ is outside of $\Delta\mu$. Suppose to a single point μ' a vector H' is assigned, so that $\eta'(\mu) = 0$ for $\mu \neq \mu'$; then H' is a proper eigen-vector of A with the eigen-value $\alpha' = \alpha(\mu')$.

There are other generalizations of the notion of spectral representation. We shall discuss them when we need them. We only mention shortly the "multiple spectral representation" in which the vectors are represented - not by a single function - but by several functions.

For example in our second problem we may write our function $\phi(s)$ in the form

$$\emptyset(s) = \sum_{\mu=0}^{\infty} \eta_1(\mu) \cos \mu s + \sum_{\mu=1}^{\infty} \eta_2(\mu) \sin \mu s$$

and thus represent them by a pair of functions $\{\eta_1(\mu), \eta_2(\mu)\}$. This representation then affords a "multiple" and "indirect" representation of the operator $A = M^2 = -d^2/ds^2$ with $\alpha(\mu) = \mu^2$. For, the function $M^2\emptyset(s)$ is given by

$$M^2\emptyset(s) = \sum_{\mu=0}^{\infty} \mu^2 \eta_2(\mu) \cos \mu s + \sum_{\mu=1}^{\infty} \mu^2 \eta_2(\mu) \sin \mu s.$$

4. Functional Calculus

One of the major achievements that may be attained with the aid of a spectral representation of an operator consists in setting up an "operational calculus".

Suppose the vector Φ in the subspace $\mathfrak{S}_A$ of the space $\mathfrak{S}$ is such that the vector $A\Phi$ is also in $\mathfrak{S}_A$; then the operator A can be applied on $A\Phi$. The operator which transforms Φ into $A(A\Phi)$ will be denoted by A^2; i.e.,

$$A^2\Phi = A(A\Phi).$$

Similarly, one can define powers A^r of A and polynomials $p(A) = \sum_{p=0}^{r} c_p A^p$.

The question arises: can one assign operators $f(A)$ to arbitrary functions $f(\alpha)$ -- or at least to functions of a large class, certainly comprising polynomials and possibly comprising continuous functions and even a variety of discontinuous functions. One says that such an assignment leads to an "operational calculus" if the following propositions hold,

(i) $\qquad f_1(\alpha) + f_2(\alpha) = f(\alpha) \Rightarrow f_1(A) + f_2(A) = f(A)$

(ii) $$f_1(\alpha)f_2(\alpha) = f(\alpha) \Rightarrow f_1(A)f_2(A) = f(A)$$

(iii) $$f_1(f_2(\alpha)) = f(a) \Rightarrow f_1(f_2(A)) = f(A).$$

Here the provision is always to be added that the operations indicated be applicable. Clearly, these propositions hold for the polynomials $p(A)$ as defined above and, conversely, whenever for an assignment of operators to functions these propositions hold, the operators assigned to polynomials are just those described above.

Suppose now the operator A admits a spectral representation, an indirect but simple one, say

$$\Phi \longleftrightarrow \{\eta(\mu)\}$$

$$A\Phi \longleftrightarrow \{\alpha(\mu)\eta(\mu)\}.$$

Then we simply define $f(A)$ as the operator which transforms the vector Φ into that vector whose representer is $f(\alpha(\mu))\eta(\mu)$:

$$f(A)\Phi \longleftrightarrow \{f(\alpha(\mu))\eta(\mu)\}.$$

The question arises for which functions $f(\alpha)$ is the operator $f(A)$ defined in this manner? In other words, we ask for which functions $f(\alpha)$ is the function $\eta_1(\mu) = f(\alpha(\mu))\eta(\mu)$ the representer of a vector Φ in $\mathfrak{S}$. This question could be answered if one knew which functions $\eta(\mu)$ are representers of vectors Φ in $\mathfrak{S}$. So far in our discussion we have left open this point; but in a later chapter a precise answer to this question will be given.

At present we can state that the definition of the operator $f(A)$ derived from a spectral representation is such that the propositions i, ii, iii hold whenever the functions $f, f_1, f_2, \ldots,$ of α involved are such that the corresponding operators are defined.

This is an immediate consequence of the fact that these propositions hold for the functions $f(\alpha(\mu))$.

5. Differential Equations

An elementary, but important, use of a spectral representation can be made in the treatment of linear differential equations.

We shall discuss three types of differential equations for vectors $\Phi = \Phi(t)$ as functions of a real variable, the "time" t. For every value $t > 0$, the vector $\Phi(t)$ is to be an element of a linear space $\mathfrak{S}$. In its dependence on the variable t, the vector $\Phi(t)$ is supposed to possess a continuous derivative $\dot{\Phi}(t)$ with respect to t.

What it means for an element of a space to depend continuously on a parameter and to be differentiable with respect to it will be explained in the next section. At present -- where we are concerned with only a preliminary discussion -- we may imagine the vectors Φ to be represented by functions; continuity and differentiability of the vector may then be understood as continuity or differentiability of these functions in any ordinary sense of these terms.

We furthermore assume that the vector $\Phi(t)$, for each $t \geq 0$, belongs to the domain $\mathfrak{S}_A$ of a (linear) operator A and satisfies one of the following three differential equations: either the "heat equation"

$$\dot{\Phi} + A\Phi = 0, \tag{5.1}$$

or the "Schroedinger equation"

$$-i\dot{\Phi} + A\Phi = 0, \tag{5.2}$$

or the "wave equation"

$$\ddot{\Phi} + A\Phi = 0. \tag{5.3}$$

The vector Φ should be prescribed for $t = 0$. The solution of the wave equation is furthermore supposed to possess a continuous second derivative and its first derivative $\dot{\Phi}$ should also be prescribed for $t = 0$.

Suppose now the operator A admits a spectral representation. For instance, let every vector Φ in $\mathfrak{S}$ be represented by a function $\eta(\mu)$ defined in an appropriate μ-domain, such that the vector $A\Phi$, for Φ in $\mathfrak{S}_A$, is represented by the function $\alpha(\mu)\eta(\mu)$.

The solution $\Phi(t)$ of any of the three differential equations is then represented by a function $\eta(\mu,t)$ of μ and t which possesses a continuous first derivative $\dot{\eta}(\mu,t)$ -- or a second derivative $\ddot{\eta}(\mu,t)$ in the case (5.3) -- and which satisfies the equation

$$\dot{\eta}(\mu,t) + \alpha(\mu)\eta(\mu,t) = 0, \tag{5.4}$$

or

$$-i\eta(\mu,t) + \alpha(\mu)\eta(\mu,t) = 0, \tag{5.5}$$

or

$$\ddot{\eta}(\mu,t) + \alpha(\mu)\eta(\mu,t) = 0, \tag{5.6}$$

respectively. Consequently, the representer $\eta(\mu,t)$ is given by

$$\eta(\mu,t) = e^{-t\alpha(\mu)}\eta(\mu,0), \tag{5.7}$$

$$\eta(\mu,t) = e^{-it\alpha(\mu)}\eta(\mu,0), \tag{5.8}$$

$$\eta(\mu,t) = \cos(t\sqrt{\alpha(\mu)})\,\eta(\mu,0) + \frac{\sin(t\sqrt{\alpha(\mu)})}{\sqrt{\alpha(\mu)}}\dot{\eta}(\mu,0), \tag{5.9}$$

respectively.

Thus the solutions of the differential equations in the η-representation are found.

If the transformation is known through which a vector Φ is given in terms of its representer $\eta(\mu)$, the solutions $\Phi(t)$ of the three equations can be determined.

These solutions can be expressed with the aid of an operational calculus in the form

$$\Phi(t) = e^{-tA}\Phi(0), \tag{5.7$_1$}$$

$$\Phi(t) = e^{-itA}\Phi(0), \tag{5.8$_1$}$$

$$\Phi(t) = \cos(t\sqrt{A})\,\Phi(0) + \frac{\sin(t\sqrt{A})}{\sqrt{A}}\dot{\Phi}(0), \tag{5.9$_1$}$$

since the operators $f(A)$ entering here are defined by virtue of the assumptions made in the preceding paragraph.

This elegant -- and helpful -- form of the solution of differential equations may serve to illustrate the striking effects that may be produced from spectral representations.

Projectors

At the end of this chapter we shall discuss a particular type of operators, the projectors, which play a dominant role in the spectral theory of operators.

Projectors are operators P for which the relation

$$P(P\Phi) = P\Phi$$

holds for all vectors Φ on which it is applicable. Using the notation of functional calculus the above relation can also be written in the form

$$P^2 = P.$$

The manifold of vectors of the form $P\Phi$ evidently forms a linear space - we denote it by $\mathfrak{P}$: the vector $P\Phi$ is called the "projection" of Φ into this space $\mathfrak{P}$. The relation $P^2 = P$ obviously expresses the condition that a vector in the space $\mathfrak{P}$ is projected into itself.

We should mention that in the literature operators P were originally called projection operators or simply projectors; we prefer to call these operators "projectors" since we like to reserve the term "projection" for the result of applying the operator.

To describe a projector in space $\mathfrak{S}$ of a finite dimension we may consider a k-dimensional subspace $\mathfrak{P}$ and an (n-k)-dimensional space $\mathfrak{P}'$ which has only the origin in common with $\mathfrak{P}$. It is known that then every vector Φ in $\mathfrak{S}$ can be written as the sum of a vector in $\mathfrak{P}$ and one in $\mathfrak{P}'$ in a unique way. Denoting these two vectors by $P\Phi$ and $(1-P)\Phi$ we realize that by virtue of their uniqueness the assignments of $P\Phi$ and $(1-P)\Phi$ to Φ constitute projectors.

In a space of functions $\phi(s)$ one may define an operator P with the aid of 2k functions $\phi_1(s), \psi_1(s), \ldots, \phi_k(s), \psi_k(s)$ through the formula

$$P\phi(s) = \sum_{\kappa=1}^{k} \phi_\kappa(s) \int \psi_\kappa(s)\phi(s)ds;$$

it is evidently a projector provided the functions ϕ_κ, ψ_κ satisfy the relations

$$\int \psi_\kappa(s)\phi_\lambda(s)ds = \delta_{\kappa\lambda}, \qquad \kappa,\lambda = 1,\ldots,k.$$

The space into which P projects consists of the linear combinations of the $\phi_1,\ldots,\phi_k$; it is thus finite dimensional.

In connection with a spectral representation of an operator A through functions $\eta(\mu)$ one should like to introduce a projector which projects a vector ϕ into the eigen-space $\mathfrak{P}_{\Delta\alpha}$ associated with an interval $\Delta\alpha$. We recall that this eigen-space consists of all those vectors whose representers $\eta(\mu)$ vanish for all values of μ for which $\alpha(\mu)$ lies outside the interval $\Delta\alpha$. Such a "spectral projector" can immediately be constructed with the aid of a functional calculus. We need only introduce the characteristic function $f_{\Delta\alpha}(\alpha)$ of the interval $\Delta\alpha$, given by

$$\begin{aligned} f_{\Delta\alpha}(\alpha) &= 1 \quad \text{for} \quad \alpha \quad \text{in} \quad \Delta\alpha \\ &= 0 \quad \text{for} \quad \alpha \quad \text{not in} \quad \Delta\alpha. \end{aligned}$$

If the operator $f_{\Delta\alpha}(A)$ can be defined for such a function it is the desired projector. For, the function $f_{\Delta\alpha}(\alpha)$ evidently satisfies the relation

$$f^2_{\Delta\alpha}(\alpha) = f_{\Delta\alpha}(\alpha);$$

hence the operator $f_{\Delta\alpha}(A)$ satisfies the relation

$$f^2_{\Delta\alpha}(A) = f_{\Delta\alpha}(A)$$

and thus is seen to be a projector. Clearly, the vector $f_{\Delta\alpha}(A)\Phi$ lies in the eigen-space $\mathfrak{P}_{\Delta\alpha}$ since its representer is $f_{\Delta\alpha}(\alpha(\mu))\eta(\mu)$ if $\eta(\mu)$ is the representer of Φ. If Φ is already in this eigen-space the relation

$$f_{\Delta\alpha}(\alpha(\mu))\eta(\mu) = \eta(\mu)$$

evidently holds and hence

$$f_{\Delta\alpha}(A)\Phi = \Phi .$$

Thus it is seen that the eigen-space $\mathfrak{P}_{\Delta\alpha}$ is exactly the space into which the "spectral projector"

$$P_{\Delta\alpha} = f_{\Delta\alpha}(A)$$

projects.

To exemplify the notions of spectral projector we consider our third example in which the functions $\phi(s)$ are represented by functions $\eta(\mu)$,

$$\phi \longleftrightarrow \{\eta(\mu)\}$$

in such a way that

$$M\phi \longleftrightarrow \{\mu\eta(\mu)\}; \text{ here } M = -id/ds.$$

From formula (1.16) we realize that the spectral projector transforms the function ϕ into the function

$$P_{\Delta\mu}\phi(s) = \int_{\Delta\mu} e^{is\mu}\eta(\mu)d\mu.$$

Substituting $\eta(\mu)$ from formula (1.16)* we find

$$P_{\Delta\mu}\phi(s) = \frac{1}{2\pi}\int_{\Delta\mu}\int_{-\infty}^{\infty} e^{i(s-s')\mu}\phi(s')ds'd\mu.$$

Thus we see that the spectral projector is given as an integral operator. This is typical for cases of operators with a continuous spectrum acting on functions of a continuous variable.

For our second example, where the spectrum is discrete, we find from (1.9) and (1.9)*, the formula

$$P_{\Delta\mu}\phi = \sum_{\mu\varepsilon\Delta\mu} \frac{1}{2\pi} \int_{-\pi}^{\pi} e^{i(s-s')\mu}\phi(s')ds'.$$

The assignment of the spectral projectors and the eigenspaces to an operator A is said to yield the "spectral resolution" of this operator. While in our presentation the spectral representation is adopted as the basic notion. We have derived the functional calculus from the spectral representation and the spectral resolution followed most frequently in the treatment of specific problems. But this procedure has disadvantages for the development of the general theory. The reason is that the spectral representation of an operator is not unique; there are many (equivalent) possibilities for it. The spectral resolution, on the other hand, is unique inasmuch as the functions $f(A)$ of an operator are uniquely assigned to it.

It is for this reason that in the development of the general theory the indicated procedure is completely reversed: first the spectral resolution - or the functional calculus - is established; a spectral representation is then derived afterwards. Thus, in the general spectral theory of bounded operators which we shall present in Chapter IV we shall in fact first establish the functional calculus and then a spectral representation.

CHAPTER II

NORM AND INNER PRODUCT

6. Normed Spaces

In order to be able to develop any specific theory such as a spectral theory in spaces of infinite dimension it is necessary to endow such a space with specific "structural" features. The requirement of linearity does not give enough structure to a space for our purposes. The central structure that we want most of our linear spaces to carry is the "inner product". Before introducing this concept we shall discuss the notion of "norm", a structural feature possessed by all linear spaces we shall deal with.

A norm associated with a linear space $\mathfrak{N}$ is a real number $||\Phi||$ assigned to every vector Φ in $\mathfrak{N}$ possessing the following properties:

$$||\Phi|| \geq 0, \tag{6.1}$$

$$||c\Phi|| = |c| \; ||\Phi||; \tag{6.2}$$

note that this last property implies

$$||0|| = 0. \tag{6.2$_o$}$$

Further properties are:

$$||\Phi|| = 0 \quad \text{implies} \quad \Phi = 0, \tag{6.3}$$

and

$$||\Phi'+\Phi|| \leq ||\Phi'|| + ||\Phi||. \tag{6.4}$$

The latter relation is the "triangle inequality". Inserting in it $\Phi' - \Phi$ in place of Φ' or $-\Phi$ and using (6.2) for $c = -1$, we obtain the "second triangle inequality"

$$\big|\, ||\Phi'|| - ||\Phi|| \,\big| \leq ||\Phi'-\Phi||. \tag{6.5}$$

If property (6.3) does not obtain we speak of a "semi-norm".

In a space of finite dimension whose vectors Ξ are represented by numbers $\xi_1,\ldots,\xi_n$ we may, for example, introduce the norm

$$||\,\Xi\,|| = \max_{\tau} |\xi_\tau|, \qquad \tau = 1,\ldots,n$$

or the norm

$$||\,\Xi\,|| = \sum_{\tau=1}^{n} |\xi_\tau|.$$

In the space of continuous $\phi(s)$, defined in an interval $\mathcal{J}$ we may, for example, assign to a function $\phi(s)$ the maximum of its absolute value as its norm

$$||\phi|| = \max_{s} |\phi(s)|, \quad s \text{ in } \mathcal{J},$$

or we may choose

$$||\phi|| = \int_{\mathcal{J}} |\phi(s)|ds.$$

All these assignments have the four properties (6.1) to (6.4) and hence qualify as norms; they would be only seminorms, if in its definition we had restricted the variable s to a subset of $\mathcal{J}$ or the variable τ to a subset of the numbers $1,\ldots,n$.

With the aid of the notion of norm we may introduce the notion of "bounded operator": a (linear) operator A defined in a subspace $\mathscr{R}_A$ and producing vectors in $\mathfrak{D}$ is called <u>bounded</u> if there is a number a such that

$$||A\Phi|| \leq a||\Phi|| \quad \text{for all } \Phi \text{ in } \mathscr{R}_A.$$

Thus an integral operator K, given by

$$K\phi(s) = \int_{\mathscr{I}} k(s,s')\phi(s')ds'$$

and acting on continuous functions $\phi(s)$ in a finite interval $\mathscr{I}$ is bounded with respect to the maximum norm $||\phi|| = \max_{s \in \mathscr{I}} |\phi(s)|$, in case the "kernel" $k(s,s')$ is a continuous function there; for, then we have

$$||K\phi|| \leq \bar{k}||\phi||$$

with $\bar{k} = \max_{s} \int_{\mathscr{I}} |k(s,s')|ds'$. Differential operators are not bounded however, as shown in Chapter VI.

We shall discuss bounded operators in great detail only in Chapter IV. Still, we have mentioned this notion already here because of its great importance and because we shall occasionally refer to it already before reaching Chapter IV.

Once a norm is established in a linear space the notion of "density" can be introduced and the notion of "dimension" can be given a more specific meaning.

A subset $\mathfrak{N}'$ of a normed space $\mathfrak{N}$ is said to be <u>dense</u> in it if to every vector Φ in $\mathfrak{N}$ a vector Φ' in $\mathfrak{N}'$ can be found such that the norm $||\Phi'-\Phi||$ of the difference, is arbitrarily small; in other words, if every Φ in $\mathfrak{N}$ can be approximated arbitrarily closely by vectors of $\mathfrak{N}'$, closeness being measured with the aid of the norm.

Thus, the space of polynomials $p(s)$ is dense in the space $\mathfrak{C}$ of continuous functions $\phi(s)$ defined in the closed interval $\mathscr{I}$ if the norm $||\phi|| = \max_{s \text{ in } \mathscr{I}} |\phi(s)|$ is adopted as the norm. That is the statement of the Weierstrass approximation theorem. The set of trigonometric polynomials $\sum_{\nu_1<\nu<\nu_2} c_\nu e^{i\nu s}$ is not dense in $\mathfrak{C}$ with respect to the maximum norm, but it is dense with respect to this norm in the subspace $\mathfrak{C}_0$, consisting of periodic functions, i.e., of functions with $\phi(\pi) = \phi(-\pi)$. But the trigonometric polynomials are dense in the space of continuous functions if the "absolute integral" norm

$$||\Phi|| = \int_{\mathscr{I}} |\phi(s)|ds,$$

is adopted.

A subset $\mathfrak{N}'$ of $\mathfrak{N}$ is said to "span" the space densely if the set $\mathfrak{N}''$ consisting of all finite linear combinations of vectors of $\mathfrak{N}'$ is dense in $\mathfrak{N}$. Thus the space of powers s^n, $n = 0,1,2,\ldots$ spans $\mathfrak{C}$ densely with respect to the maximum norm.

In Section 2 we said that a space is of infinite dimension if it does not consist of the linear combinations of a finite number of vectors. We now say that a space is of "countable" or "denumerable dimension" if there is a countable set, i.e., an infinite sequence of vectors which spans the space densely. Thus the space of continuous functions mentioned is of countable dimension with respect to the maximum norm. Note that the notion of dimension, as introduced here, depends on the choice of the norm. In fact, as we shall show in a later section, the same space can be made to have nondenumerable dimension by imposing another norm on it.

Spaces of countable dimension are more customarily called "separable".

Note that the space of continuous functions $\phi(s)$ is of countable dimension, or separable, with respect to the maximum norm and the absolute integral norm, in spite of the fact that the vectors of this space possess a non-countable number of components - if we consider as these components the values of $\phi(s)$ at all the points s.

7. Inner Product

The inner product, the central structural feature of the linear spaces we shall investigate, is the natural analogue of the inner product of two vectors in Euclidean geometry. The natural analogue of the magnitude of an Euclidean vector is a norm which can be formed with the aid of an inner product.

Solely because of the analogy with Euclidean geometry, the geometry of spaces carrying an inner product would be worthwhile to investigate. For us there is a more specific reason for doing so. With reference to an inner product the notion of "self-adjointness" of an operator can be defined. Self-adjoint operators have a remarkable property: they admit a spectral representation and their spectra are real. Self-adjoint operators are the primary object of this set of notes.

The inner product associated with a linear space $\mathfrak{B}$ is a complex number (Φ',Φ) assigned to any pair Φ',Φ of vectors in $\mathfrak{B}$ possessing the following properties, (7.1) to (7.4):

It is linear in the second "factor":

$$(\Phi',c_1\Phi^{(1)} + c_2\Phi^{(2)}) = c_1(\Phi',\Phi^{(1)}) + c_2(\Phi',\Phi^{(2)}). \tag{7.1}$$

It is symmetric in the Hermitean sense:

$$(7.2) \qquad (\Phi,\Phi') = \overline{(\Phi',\Phi)};$$

the bar here indicates that the complex conjugate is to be taken. Property (7.2) implies that the associated quadratic form (Φ,Φ) is real. With this fact in mind we can formulate the next property of an inner product: The quadratic form (Φ,Φ) is positive definite. This property will be split into two parts:

$$(7.3) \qquad (\Phi,\Phi) \geq 0,$$

and

$$(7.4) \qquad (\Phi,\Phi) = 0 \quad \text{implies} \quad \Phi = 0.$$

A space in which an inner product is defined which has these four properties is called an inner-product-space; if it has the first three properties but not necessarily the last one, (7.4), it will be called a semi-inner-product-space.

If the space $\mathfrak{V}$ is real, the inner product is linear also in the first factor; if it is complex, the relation

$$(7.1)^* \qquad (c_1\Phi^{(1)} + c_2\Phi^{(2)},\Phi') = \bar{c}_1(\Phi^{(1)},\Phi') + \bar{c}_2(\Phi^{(2)},\Phi')$$

holds, as follows from (7.1) combined with (7.2). One refers to this relation by saying that the inner product is "antilinear" in the first factor. (The term conjugate linear would seem more appropriate.)

The most important basic property of such an inner product space is embodied in "Schwarz's inequality"

$$(7.5) \qquad |\Phi',\Phi|^2 \leq (\Phi',\Phi')(\Phi,\Phi).$$

For convenience we have written $|\Phi',\Phi|$ for the absolute value of the inner product (Φ',Φ) instead of $|(\Phi,\Phi')|$. To prove this inequality we first assume that the inner product (Φ',Φ) is real. With two arbitrary real numbers a, a_1 we then derive from (7.1), (7.1)* the relation

$$
\begin{aligned}
0 &\leq (a\Phi + a_1\Phi',\ a\Phi + a_1\Phi') \\
&= a^2(\Phi,\Phi) + 2aa_1(\Phi',\Phi) + a_1^2(\Phi',\Phi'),
\end{aligned}
$$

which expresses the fact that the right member here is a non-negative quadratic form in a, a_1. Therefore (7.5) holds. If (Φ',Φ) is not real, we let θ be a number of absolute value 1 such that $\theta(\Phi',\Phi)$ is real and non-negative, i.e.

$$\theta(\Phi',\Phi) = |\Phi',\Phi| \geq 0.$$

Then $(\Phi',\theta\Phi) = \theta(\Phi',\Phi)$ is non-negative, and hence the statement results from

$$|\Phi',\Phi|^2 = (\Phi',\theta\Phi)^2 \leq (\Phi',\Phi')(\theta\Phi,\theta\Phi) = (\Phi',\Phi')(\Phi,\Phi).$$

Note that the property (7.4) was not used in this proof. Consequently, the Schwarz inequality holds also for semi-inner-product-spaces.

Furthermore, one readily verifies that equality holds in the Schwarz inequality if and only if there are two complex numbers c and c_1, not both being zero, such that $c\Phi + c_1\Phi' = 0$ - or $|c\Phi+c_1\Phi',\ c\Phi + c_1\Phi'|$ in case of a semi-inner-product-space.

The quadratic form (Φ,Φ) will be called the "unit form". We maintain that -- as in Euclidean geometry -- the square root of the unit form

(7.6) $$||\Phi|| = \sqrt{(\Phi,\Phi)}$$

may serve as a norm. The fact that (Φ,Φ) is real follows from (7.2) as noted above. Properties (6.1) and (6.2) are immediate consequences of (7.3) and (7.1). Property (6.3) is implied by (7.4); if (7.3) does not obtain, $||\Phi||$ is a semi-norm. It remains to prove property (6.4), the triangle inequality.

This inequality is an immediate consequence of the Schwarz inequality. We first write this inequality in the form

(7.5)' $$|\Phi',\Phi| \leq ||\Phi'||\ ||\Phi||$$

and then proceed as follows:

$$\begin{aligned} ||\Phi+\Phi'||^2 &= (\Phi+\Phi',\Phi+\Phi') \\ &= (\Phi,\Phi) + (\Phi,\Phi') + (\Phi',\Phi) + (\Phi',\Phi') \\ &\leq ||\Phi||^2 + ||\Phi||\ ||\Phi'|| + ||\Phi'||\ ||\Phi|| + ||\Phi||^2 \\ &= \{||\Phi|| + ||\Phi'||\}^2; \end{aligned}$$

this is the statement.

It is clear from this proof and the remarks made at the end of the proof of Schwarz's inequality that equality holds in the triangle inequality if and only if there are two complex numbers c, c_1 such that $c\Phi + c_1\Phi' = 0$.

8. Inner Products in Function Spaces

We proceed to discuss various specific expressions for inner products commonly adopted in specific linear spaces commonly considered. In doing this we shall frequently -- for convenience -- just describe the unit form; the proper expression of the inner product can be derived from it in an obvious manner.

In a real finite dimensional space of vectors Ξ represented by n components $\xi_1,\ldots,\xi_n$ the commonly adopted inner product is the one associated with Euclidean geometry

$$\left(\Xi',\Xi\right) = \sum_{\sigma=1}^{n} \xi'_\sigma \xi_\sigma .$$

In a complex finite-dimensional space one commonly adopts the analogous Hermitean inner product

$$(8.1) \qquad \left(\Xi',\Xi\right) = \sum_{\sigma=1}^{n} \bar{\xi}'_\sigma \xi_\sigma .$$

The validity of the requirements (7.1) to (7.4) is immediately verified.

Of course, other bilinear forms associated with positive definite quadratic forms could be chosen.

As an example of an infinite-dimensional space carrying an inner product we consider the space of continuous functions $\phi(s)$ defined in an interval $\mathscr{R}$ of the s-axis; for these functions we may define as inner product the integral

$$(8.2) \qquad (\phi',\phi) = \int_{\mathscr{R}} \overline{\phi'(s)}\phi(s)\,ds.$$

Clearly, requirements (7.1) - (7.4) are satisfied.

The associated norm in this inner product function space is evidently

$$(8.2)' \qquad ||\phi|| = \left[\int_{\mathscr{R}} |\phi(s)|^2 ds\right]^{1/2}.$$

Instead of a finite interval we may take for $\mathcal{R}$ a ray $s_o \leq s < \infty$ or the full s-axis, $-\infty < s < \infty$. In that case we require at present that each function $\phi(s)$ be of bounded support, i.e., vanishes identically outside of an appropriate finite interval. For such functions the inner product (8.2) is defined. Later on we shall enlarge this class of functions.

Our aims in this chapter, as in Chapter I, are to present the formal aspects of spectral representation. In Chapter III and subsequent chapters we shall extend the classes of admitted functions so as to attain the desired completeness of the theory.

Accordingly, we may suit ourselves in the choice of the class of functions $\phi(s)$, once the region $\mathcal{R}$ and the inner product have been selected. On the other hand, we should already at the present state attain a certain generality in the choice of the region $\mathcal{R}$ and the inner product.

Before describing such generalizations it is opportune to enlarge the class of continuous functions to the class of "piece-wise continuous" functions, and to employ a more general notion of integration, "integration with respect to a measure function".

First, we introduce the notion of partition of the s-axis. As a partition $\mathcal{P}$ we consider an alternating sequence of open intervals and points:

$$\mathcal{I}_\sigma: \; s_{\sigma-1} < s < s_{\sigma+1}, \qquad \sigma \text{ even},$$
$$\mathcal{I}_\sigma: \qquad s = s_\sigma \qquad , \qquad \sigma \text{ odd},$$
$$-\infty < \sigma < \infty.$$

We prefer this type of partition to other types of partitions which could, of course, also be used.

We call a function $f(s)$ "piecewise constant" if, with

reference to an appropriate partition,

$$f(s) = f_{\sigma} = \text{constant for } s \text{ in } \mathscr{I}_{\sigma}.$$

Clearly, such a function is also piecewise constant with reference to a subdivision of $\mathscr{P}$, obtained by subdividing some or all of the open intervals of $\mathscr{P}$ again into open intervals and points. Since two partitions evidently have a common subdivision, it is clear that piecewise constant functions form a linear space.

Every continuous function can be approximated by piecewise constant functions, uniformly in every bounded interval. We might enlarge the space $\mathfrak{C}$ of all continuous functions to the space of all functions which can be approximated by piecewise constant ones, uniformly in every bounded interval. We are satisfied, however, with considering a restricted extension of the space $\mathfrak{C}$: the space $\mathfrak{C}'$ of "piecewise continuous functions". We call a function $f(s)$ piecewise continuous if, in each open interval $s_{\sigma-1} < s < s_{\sigma+1}$ of a partition, $f(s)$ agrees with a function which is continuous in the closure, $s_{\sigma-1} \leq s \leq s_{\sigma+1}$, of this interval. (The values of this continuous function at the end points need not be related to the values of $f(s)$ there.) Clearly, each piecewise continuous function can be approximated by piecewise constant ones, uniformly in every bounded interval.

Piecewise continuous, as well as continuous, functions evidently form linear spaces, and the product of two functions of any of these spaces again belongs to that space.

To attain a more general form of an inner product in a function space we employ the notion of integration with respect to a "measure function". Such a function $r(s)$ is a non-decreasing real-valued function defined for all values of s; i.e., for $-\infty < s < \infty$. The values of $r(s)$ at points s where $r(s)$ is not continuous are irrelevant in the following; therefore we introduce at each point s the limit values $r^+(s)$ and $r^-(s)$ which the values $r(s')$ approach when s' approaches s from above and below respectively. We call $r^{\pm}(s)$ a "measure function pair".

To each interval $\mathscr{I}_\sigma$ of a partition $\mathscr{P}$ we now assign as its measure the difference

$$\Delta_\sigma r = r^-(s_{\sigma+1}) - r^+(s_{\sigma-1}) \quad \text{for } \sigma \text{ even,}$$

i.e., if $\mathscr{I}_\sigma$ is an open interval, and

$$\Delta_\sigma r = r^+(s_\sigma) - r^-(s_\sigma) \quad \text{for } \sigma \text{ odd,}$$

i.e., if $\mathscr{I}_\sigma$ is a point. Clearly, if the open interval $\mathscr{I}_\sigma$ is subdivided

$$\mathscr{I}_\sigma = \sum_\tau \mathscr{I}'_\tau ,$$

we have

$$\Delta_\sigma r = \sum_\tau \Delta'_\tau r ,$$

in obvious notation.

Let $\mathscr{R}$ be an open or closed interval; then for any piecewise constant function $f(s)$ - with reference to a partition which contains the end points of $\mathscr{R}$ - we define the integral of $f(s)$ with respect to r by

$$\int_{\mathscr{R}} f(s)\,dr(s) = \sum_{\mathscr{I}_\sigma \text{ in } \mathscr{R}} f_\sigma \Delta_\sigma r .$$

Clearly, this value is not changed by subdivision of $\mathscr{P}$.

If now $f(s)$ is a function in the space $\mathfrak{C}'$ we may define its integral simply as the limit

$$\int_{\mathscr{R}} f(s)dr(s) = \lim_{\nu \to \infty} \int_{\mathscr{R}} f^{(\nu)}(s)dr(s)$$

of uniformly approximating piecewise constant functions $f^{(\nu)}(s)$. From the inequality

$$\left| \int_{\mathscr{R}} f(s)dr(s) \right| \leq \max |f(s)| \int_{\mathscr{R}} dr(s),$$

it follows that this limit is independent of the choice of the sequence of approximating functions $f^{(\nu)}(s)$.

If $f(s) \geq 0$, obviously $\int_{\mathscr{R}} f(s)dr(s) \geq 0$. On the other hand, the relation

$$\int_{\mathscr{R}} f(s)dr(s) = 0 \quad \text{together with} \quad f(s) \geq 0$$

does not imply $f(s) \equiv 0$ unless special provisions are made.

First of all the values of $f(s)$ in any open interval in which $r(s)$ is constant do not contribute to the value of the integral. Therefore, we may simply consider the function $f(s)$ as defined only in the "carrier" of the measure function $r(s)$, i.e., the set obtained by removing all open constancy intervals of $r(s)$. Still, even the integral $\int_{\mathscr{R}} f(s)dr(s)$ may be zero for a function $f(s)$ which is ≥ 0 but $\not\equiv 0$. For if the value of $f(s)$ is changed at a point s at which $r(s)$ is continuous, the value of the integral is not changed. If, however, we require that $f(s)$ should be continuous where $r(s)$ is continuous, the statement that

$$\int_{\mathscr{R}} f(s)dr(s) = 0 \quad \text{for} \quad f(s) \geq 0 \quad \text{implies} \quad f(s) = 0$$

becomes valid, as easily verified.

In the space of $\mathfrak{C}'_{\mathscr{R}}$ functions $\phi(s)$ in $\mathfrak{C}'$ which are defined in the interval $\mathscr{R}$ we may adopt the expression

$$(\phi,\phi') = \int_{\mathscr{R}} \phi(s)\phi'(s)dr(s)$$

as semi-inner product. The properties (7.1), (7.2), and (7.3) required of an inner product are verified from the fact that they obtain for piecewise constant functions. If we consider the function $\phi(s)$ as defined only in the carrier $\mathscr{R}'$ of $r(s)$ inside $\mathscr{R}$ and stipulate that it be continuous where $r(s)$ is, the inner product is a strict one, i.e., (7.4) holds, since then

$$\int_{\mathscr{R}} |\phi(s)|^2 dr(s) = 0 \quad \text{implies} \quad \phi(s) = 0.$$

We may also consider the space $\dot{\mathfrak{C}}'$ consisting of functions $\phi(s)$ in $\mathfrak{C}'$ which vanish identically outside a bounded interval (depending on ϕ), and set

$$(\phi,\phi') = \int \phi(s)\phi'(s)dr(s)$$

where the integration is to be extended over any interval outside of which $\phi(s) = 0$.

Such functions will be said to have a "bounded support", the support of a function being the closure of the set on which it does not vanish. An extension to spaces of functions which are not required to be of bounded support will be given in Chapter III.

To illustrate the variety of cases in which the inner product is given by an integral of the kind described let us first take the function

$$
\begin{aligned}
r(s) &= 0 \qquad , \quad s < 0 \\
&= \rho + s \quad , \quad 0 \leq s \leq s_1 \\
&= \rho + s_1 \quad , \quad s_1 \leq s \quad ,
\end{aligned}
$$

for $\rho > 0$. The functions $\phi(s)$ may be defined as zero for $s < 0$ and $s > s_1$. The unit form, i.e., the inner product of ϕ with itself, can be written as

$$
\int_0^{s_1} |\phi(s)|^2 ds + \rho|\phi(0)|^2 \ ;
$$

note that the value $\phi(0)$ is an independent quantity not related by a continuity requirement to the values of $\phi(s)$ for $s > 0$.

Generalizing further, we may consider the unit form to be given by

$$
\sum_{\nu=1}^{\infty} \rho_\nu |\phi(s_\nu)|^2 + \int_0^{\infty} |\phi(s)|^2 ds,
$$

where $s_1, s_2, \ldots$ is an increasing sequence of negative numbers, and ρ_1, ρ_2 is a sequence of positive numbers with $\sum_{\nu=1}^{\infty} \rho_\nu < \infty$. We realize easily that this expression could be written as an integral with respect to a measure function.

Finally, we realize that a unit form given as a sum

$$
\sum_{\nu=1}^{\infty} \rho_\nu |\phi(s_\nu)|^2
$$

may be written as such an integral if the s_ν are any sequence of distinct numbers and the positive numbers ρ_ν are only required to be such that their sum converges, $\sum_{\nu=1}^{\infty} \rho_\nu < \infty$. As a matter of fact, the numbers s_ν might form a dense set, e.g., they might consist of all rational numbers in a finite interval.

All such possibilities are important in spectral theory.

Naturally, we may also consider functions $\phi(s)$ of a multiple variable $s = s_1, \ldots, s_k$ defined in a region $\mathscr{R}$ of the k-dimensional space and set up inner products with the unit form

$$(\phi,\phi) = \int_{\mathscr{R}} |\phi(s)|^2 ds,$$

where $ds = ds_1 \cdots ds_k$. The notion of integral with respect to a measure could, of course, be extended to functions of several variables.

We may even consider functions $\phi(s)$ of an "infinite" variable $s = s_1, s_2, \ldots$. In that case the notion of integral of functions of infinitely many variables will have to be employed. We may go even further and let s stand for a continuum of variables, for example by letting s stand for the values of a function $s(\xi)$ of a continuous variable ξ.

The latter possibilities are of importance in spectral problems of the quantum theory of fields. We shall discuss such problems in detail at the end of these notes.

9. Formally Self-Adjoint Operators

At the beginning of Section 7 it was said that the important property of self-adjointness of an operator refers to an inner product defined in space $\mathfrak{B}$.

Suppose A and $A^\dagger$ are two operators, defined in subspaces $\mathfrak{B}_A$ and $\mathfrak{B}_{A^\dagger}$ of $\mathfrak{B}$, which have the property that the relation

$$(9.1) \qquad (\Phi', A^\dagger\Phi) = (A\Phi', \Phi)$$

holds for all vectors Φ in $\mathfrak{B}_{A^\dagger}$, Φ' in $\mathfrak{B}_A$. Then we say that the operators A and $A^\dagger$ are "formally adjoint" to each other. If the operator A is formally adjoint to itself, so that $A^\dagger = A$ and the relation

$$(9.2) \qquad (\Phi',A\Phi) = (A\Phi',\Phi)$$

holds for all vectors Φ,Φ' in $\mathfrak{B}_A$, the operator A will be called "formally self-adjoint".

The qualification "formal" is employed in order to distinguish "formal" adjointness from the more restrictive property of "strict" adjointness, which we shall discuss in later parts of these notes.

The notion of formal self-adjointness is the natural generalization of the notion of symmetry of a matrix; in fact, if the operator acts in a finite-dimensional real space, the formal self-adjointness just reduces to the symmetry of the representing matrix. The formal adjoint of an operator then corresponds to the transposed matrix.

To see this consider a finite dimensional real space of vectors Ξ given by sets of numbers $\{\xi_1,\ldots,\xi_n\}$ and suppose the inner product of two vectors Ξ and Ξ' is given by

$$\left(\Xi',\Xi\right) = \sum_{\sigma=1}^{n} \xi'_\sigma \xi_\sigma .$$

Let the operators A and $A^\dagger$ be represented by matrices $\{a_{\sigma\sigma'}\}$ and $\{a^\dagger_{\sigma\sigma'}\}$, so that

$$\left(\Xi',A^\dagger\Xi\right) = \sum_{\sigma,\sigma'=1}^{n} \xi'_\sigma a^\dagger_{\sigma\sigma'} \xi_{\sigma'} ,$$

$$\left(A\Xi',\Xi\right) = \sum_{\sigma,\sigma'=1}^{n} a_{\sigma\sigma'} \xi'_{\sigma'} \xi_\sigma .$$

The identity of these expressions for all values of $\{\xi_\sigma\}$ and $\{\xi'_\sigma\}$

evidently leads to the relation

$$a^{\dagger}_{\sigma\sigma'} = a_{\sigma\,\sigma} \quad ,$$

which shows that the matrix $\{a^{\dagger}\}$ is indeed the transpose of $\{a\}$. Formal self-adjointness, $A^{\dagger} = A$, is thus seen to be equivalent to the symmetry $a_{\sigma\,\sigma'} = a_{\sigma'\,\sigma}$ of $\{a\}$.

If the finite dimensional space is complex and carries the inner product

$$\left(\Xi', \Xi\right) = \sum_{\sigma=1}^{n} \bar{\xi}'_{\sigma}\xi_{\sigma} \quad ,$$

the formal adjointness relation

$$\sum_{\sigma,\sigma'=1}^{n} \bar{a}_{\sigma\sigma'}\bar{\xi}'_{\sigma'}\xi_{\sigma} = \sum_{\sigma,\sigma'=1}^{n} \bar{\xi}'_{\sigma}a^{\dagger}_{\sigma\sigma'}\xi'_{\sigma}$$

between the operators A and $A^{\dagger}$ leads to the relation

$$\overline{a_{\sigma\,\sigma'}} = a^{\dagger}_{\sigma'\sigma} \quad ,$$

which shows that the matrix $\{a^{\dagger}\}$ is the "Hermitean transpose" of $\{a\}$. Formal self-adjointness, $A^{\dagger} = A$, then is equivalent with "Hermitean" symmetry of $\{a\}$

$$\overline{a_{\sigma\,\sigma'}} = a_{\sigma'\sigma} \quad .$$

In view of these relationships, formally self-adjoint operators in any inner product space are also called symmetric if the space is real and Hermitean if it is complex.

We recall that the matrix we dealt with in Example 1

in Section 2 was required to be symmetric. It is a well known consequence of this symmetry that a principal axes transformation is possible and that the eigenvalues are real.

In the present chapter we shall not yet be able to produce a spectral representation -- the analogue of the principal axes transformation -- for operators acting in spaces of infinite dimension. We are, however, able to establish the other basic property mentioned, namely the property of formally self-adjoint operators that their eigenvalues are real.

This fact is a consequence of the important property of any formally self-adjoint operator A, that the quadratic form $(\Phi,A\Phi)$ associated with it assumes only real values. That this is so follows from formulas (7.2) and (9.2) which yield the relation

$$\overline{(\Phi,A\Phi)} = (A\Phi,\Phi) = (\Phi,A\Phi).$$

To show that the point eigenvalues of a formally self-adjoint operator are real let α be such an eigenvalue associated with the eigenvector $H \neq 0$, so that $AH = \alpha H$. This eigenvalue can evidently be expressed as the quotient

$$\alpha = \frac{(H,AH)}{(H,H)} \tag{9.3}$$

of the forms (H,AH) and (H,H). Since the values of these forms are real, α is real.

We shall not attempt to prove an analogous statement for continuous spectra of formally self-adjoint operators. For a strictly "self-adjoint operator", the reality of its spectrum will be evident from its spectral representation, see Chapters IV and VI. In these chapters we shall also describe the form the inner product assumes if the vectors of the space are given in spectral representation.

10. Adjoint Operators in Function Spaces

In the present section we shall discuss the meaning of formal adjointness for operators acting in "function spaces".

Suppose that the space $\mathfrak{B}$ consists of piecewise continuous functions $\phi(s)$ defined in the carrier $\mathscr{R}'$ of a measure function $r(s)$ in an interval $\mathscr{R}$ and suppose the inner product is given by

$$(\phi',\phi) = \int_{\mathscr{R}} \overline{\phi'(s)}\phi(s)\,dr(s).$$

Suppose further that the operators A and B are given as integral operators with kernels $\alpha(s,s')$, $\beta(s,s')$ defined when s and s' are in the carrier $\mathscr{R}'$ (at present, we assume these functions to be continuous there):

$$\text{(10.1)}\qquad \begin{aligned} A\phi(s) &= \int_{\mathscr{R}'} \alpha(s,s')\phi(s')\,dr(s') \\ B\phi(s) &= \int_{\mathscr{R}'} \beta(s,s')\phi(s')\,dr(s'). \end{aligned}$$

Here the terms $A\phi(s)$, $B\phi(s)$ stand for the values of the function $A\phi$ and $B\phi$ at the point (s). Then the operator B is formally adjoint to A exactly if

$$\text{(10.2)}\qquad \beta(s,s') = \overline{\alpha(s',s)}.$$

The integral operator is formally self-adjoint, or Hermitean, $A^{\dagger} = A$, if its kernel $\alpha(s,s')$ satisfies the relation

$$\text{(10.3)}\qquad \alpha(s,s') = \overline{\alpha(s',s)}.$$

This condition on the kernel of an integral operator is quite analogous to the condition of Hermitean symmetry of a matrix.

For differential operators the analogy is less obvious.

Let us specifically consider the operator

$$D = \frac{d}{ds}$$

acting on functions $\phi(s)$ defined in a region $\mathscr{R}$, which may be a finite closed interval of the s-axis, a ray, such as $0 \le s < \infty$, or the total s-axis.

We require that the functions $\phi(s)$ are at least continuous in $\mathscr{R}$; if the domain extends to infinity, we require the functions to be of bounded support; i.e., each function $\phi(s)$ should be identically zero outside an appropriate finite interval (depending on ϕ). Then certainly

$$(\phi,\phi) = \int_{\mathscr{R}} |\phi(s)|^2 ds$$

is finite.

The operator D will be applicable only in a subspace of the space $\mathfrak{B} = \mathfrak{C}$ of continuous functions with bounded support just described. Let us denote by $\mathfrak{C}_1$ the space of functions $\phi(s)$ in $\mathfrak{C}_1$ which have a continuous derivative $\phi(s) = d\phi(s)/ds$. Then D is defined in $\mathfrak{B}_A = \mathfrak{C}_1$.

In case the interval $\mathscr{R}$ is the total s-axis the operator $-D$ is formally adjoint to D; for, we have

$$\begin{aligned}(\phi',D\phi) + (D\phi',\phi) &= \int_{-\infty}^{\infty} \overline{\phi'(s)}\dot{\phi}(s)ds + \int_{-\infty}^{\infty} \overline{\dot{\phi}(s)}\phi(s)ds \\ &= \int_{-\infty}^{\infty} \frac{d}{ds}(\overline{\phi'(s)}\phi(s))ds = 0\end{aligned}$$

since both $\phi(s)$ and $\phi(s')$ were required to vanish identically outside appropriate finite intervals. The same statement is not true if the domain $\mathscr{R}$ has one or two finite endpoints.

Still, for such an operator D we can find a formally adjoint operator. It will be defined in the domain $\mathring{\mathfrak{C}}_1$ of all functions $\phi(s)$ in $\mathfrak{C}_1$, which vanish at the finite endpoints of $\mathscr{R}$. The operator D when restricted to this domain $\mathring{\mathfrak{C}}_1$ will be called $\mathring{D}$. Clearly, we may state that $D^\dagger = -\mathring{D}$ is formally adjoint to D:

$$(\phi', D\phi) + (\mathring{D}\phi', \phi) = 0 \quad \text{for} \quad \phi \quad \text{in} \quad \mathfrak{C}_1, \phi' \quad \text{in} \quad \mathring{\mathfrak{C}}_1.$$

since the left member here equals the expression

$$\int_{\mathscr{R}} \frac{d}{ds} (\overline{\phi'(s)}\phi(s))ds,$$

which vanishes since $\phi'(s)$ vanishes at the endpoints of $\mathscr{R}$.

If we restrict also the function ϕ to belong to the subspace $\mathring{\mathfrak{C}}_1$ the above relation remains true:

$$(\phi', \mathring{D}\phi) + (\mathring{D}\phi', \phi) = 0 \quad \text{for} \quad \phi, \phi' \quad \text{in} \quad \mathring{\mathfrak{C}}_1,$$

so that $-\mathring{D}$ is also formally adjoint to $\mathring{D}$. This fact is equivalent with the statement that the operator $M = i\mathring{D}$ is formally self-adjoint:

$$(\phi', M\phi) = (M\phi', \phi) \quad \text{for} \quad \phi, \phi' \quad \text{in} \quad \mathring{\mathfrak{C}}_1.$$

For, we have

$$(\phi', M\phi) - (M\phi', \phi) = i(\phi', \mathring{D}\phi) + i(\mathring{D}\phi', \phi) = 0.$$

This self-adjointness was achieved by restricting the domain $\overset{\circ}{\mathfrak{C}}_1$ of applicability of d/ds to the domain $\mathfrak{C}_1$. One may wonder whether this restriction was not too severe. As will be shown in Chapter VI, no spectral representation of the operator M in $\overset{\circ}{\mathfrak{C}}_1$ is possible and at the same time it will be shown that no extension of its domain will help if the region $\mathscr{R}$ has only one finite endpoint.

With the aid of the operators D and $\overset{\circ}{D}$ we may form differential operators of the second order. Let us denote by $\mathfrak{C}_L$ the space of those functions $\phi(s)$ in $\overset{\circ}{\mathfrak{C}}_1$ for which $\overset{\circ}{D}\phi = \frac{d}{ds}\phi(s)$ is in $\mathfrak{C}_1$. Then we set

$$L = -D\overset{\circ}{D} = -\frac{d^2}{ds^2} . \tag{10.4}$$

Clearly, this operator L, defined in $\mathfrak{C}_L$, is Hermitean. For, with ϕ and ϕ' in $\mathfrak{C}_L$ we have

$$(\phi',L\phi) = -(\phi',D\overset{\circ}{D}\phi) = (\overset{\circ}{D}\phi',\overset{\circ}{D}\phi)$$

and just as well

$$(L\phi',\phi) = -(D\overset{\circ}{D}\phi',\phi) = (\overset{\circ}{D}\phi',\overset{\circ}{D}\phi)$$

since D and $\overset{\circ}{D}$ are formally adjoint.

Note that the boundary condition which restricts D to $\overset{\circ}{D}$ affects only one factor $\frac{d}{ds}$ of $(\frac{d}{ds})^2$. It is for this reason that the operator $-D\overset{\circ}{D}$, (after its domain has been properly extended) can be made into a strictly self-adjoint one; see Chapter VI.

The condition that the function $\phi(s)$ on which the operator $-d^2/ds^2$ acts should vanish at the endpoints is called the "first" boundary condition. The second boundary condition on the domain of

$-d^2/ds^2$, which requires the derivative of ϕ to vanish at the endpoints, can be handled in a similar manner. One need only interpret the operator $-d^2/ds^2$ as the operator $-\overset{\circ}{D}D$ applicable on those functions $\phi(s)$ in $\mathfrak{C}_1$ for which the derivative $D = d/ds$ is in $\overset{\circ}{\mathfrak{C}}_1$ and vanishes at the endpoints. One immediately verifies that this operator is also Hermitean.

11. Orthogonality

Two vectors Φ, Φ' are said to be orthogonal or perpendicular to each other if their inner product is zero:

$$(\Phi, \Phi') = 0. \tag{11.1}$$

Evidently, this terminology is chosen in analogy with that used in Euclidean geometry. Just as in Euclidean geometry, the notion of "orthogonal coordinate system" is of importance for inner product spaces. We say, a finite or infinite system of vectors $\Omega^{(1)}$, $\Omega^{(2)}, \ldots,$ is "orthonormal" if

$$\begin{aligned} (\Omega^{(\kappa)}, \Omega^{(\kappa)}) &= 1, \qquad \kappa = 1,2,\ldots\,, \\ (\Omega^{(\kappa)}, \Omega^{(\lambda)}) &= 0, \qquad \kappa \neq \lambda = 1,2,\ldots\,. \end{aligned} \tag{11.2}$$

A finite linear combination of such vectors $\Omega^{(\kappa)}$ is a vector which can be written as a sum

$$\Phi = \sum_{\kappa=1,2,\ldots} c_\kappa \Omega^{(\kappa)} \tag{11.3}$$

in which only a finite number of coefficients c_κ differ from zero. These coefficients can be retrieved with the aid of the formula

(11.4) $$c_\kappa = (\Omega^{(\kappa)}, \Phi),$$

which is immediately verified.

The linear space $[\Omega^{(\kappa)}]$ formed by all these finite combinations will be said to be "spanned" by the system of vectors $\Omega^{(\kappa)}$.

We observe that the spectral representations in our first two examples involved such an orthonormal system; the unit vectors Ω were eigen-vectors H of the operator considered. That was no accident. Quite generally, the following theorem holds:

Theorem 11.1. Eigen-vectors of a Hermitean operator which belong to two distinct eigen-values are orthogonal.

Let A be the operator, $H^{(1)} \neq 0$, $H^{(2)} \neq 0$, be the eigen-vectors, $\alpha_1 \neq \alpha_2$ the eigen-values. Then we have

$$\alpha_2(H^{(1)}, H^{(2)}) = (H^{(1)}, \alpha_2 H^{(2)}) = (H^{(1)}, AH^{(2)})$$
$$= (AH^{(1)}, H^{(2)}) = \alpha_1(H^{(1)}, H^{(2)}).$$

Here we have made use of the fact, proved in Section 9, that the eigen-values of a Hermitean operator are real. Since $\alpha_1 \neq \alpha_2$ was assumed, the relation $(H^{(1)}, H^{(2)}) = 0$, and thus the statement of the theorem, follows.

12. Orthogonal Projection

Spectral representation of operators with a continuous (or partly continuous) spectrum cannot be described simply with the aid of orthogonal systems of eigen-vectors in the proper sense; eigen-vectors assigned to intervals in the spectrum may be used. In this connection the notion of "projection" plays a leading part.

In Section 5 we defined a projector as a linear operator P

for which $P^2 = P$; such a projector projects every vector on which it is applicable into the space $\mathfrak{B}$ of vectors of the form $P\Phi$, in such a way that every vector in this space is transformed into itself.

A projector acting in an inner-product-space $\mathfrak{B}$ may have the property that the difference $\Phi - P\Phi$ of any vector Φ and its projection $P\Phi$ is orthogonal to the space $\mathfrak{P}$ into which P projects. Such a projector is called an "orthogonal projector". From the property

$$(12.1) \qquad (P\Phi,(1-P)\Phi') = 0 \quad \text{for all} \quad \Phi,\Phi' \qquad (\text{admitting } P)$$

we deduce the important fact that an <u>orthogonal projector is formally self-adjoint</u>. For, relation (12.1) together with

$$((1-P)\Phi,P\Phi') = 0$$

leads to the relation

$$(12.2) \qquad (P\Phi,\Phi') = (\Phi,P\Phi') \quad \text{for all} \quad \Phi,\Phi'$$

admitting P.

Conversely, every Hermitean projector is orthogonal. For, relation (12.1) follows from (12.2) by substituting $P\Phi$ for Φ.

Writing the vector Φ in the form

$$(12.3) \qquad \Phi = P\Phi + (1-P)\Phi$$

and evaluating $||\Phi||^2$ by using (7.1) and (7.1)* we obtain as immediate consequence of relation (12.1) the identity

$$||\Phi||^2 = ||P\Phi||^2 + ||(1-P)\Phi||^2 \tag{12.4}$$

which may be regarded as an extension of the Pythagorean theorem.

We denote by $\mathfrak{P}^{\perp}$ the "complementary space" of $\mathfrak{P}$ i.e., the space of all vectors orthogonal to $\mathfrak{P}$. In view of (12.1) we then may interpret formula (12.3) by saying that every vector Φ admitting P can be written as the sum of a vector in $\mathfrak{P}$ and one in $\mathfrak{P}^{\perp}$.

From identity (12.4) we infer the inequality

$$||P\Phi|| \leq ||\Phi|| \tag{12.4'}$$

which expresses the fact that an orthogonal projection of a vector is shorter than the vector itself. In particular, this formula implies that an orthogonal projector is bounded in the sense explained at the end of Section 6.

Projection on a one-dimensional space is always possible. Let this space $\mathfrak{P}_1$ consist of vectors of the form $c\Omega$, where $\Omega \neq 0$. Then the projection $P_1\Phi$ on this space is given by

$$P_1\Phi = \frac{(\Omega,\Phi)}{(\Omega,\Omega)}\,\Omega \tag{12.5}$$

as is immediately verified. Projection into a space $\mathfrak{P}_n$ of finite dimension is also always possible. Let this space consist of the linear combinations of vectors $\Omega^{(1)},\ldots,\Omega^{(n)}$. Then the projection

$$P_n\Phi = \xi_1\Omega^{(1)} + \cdots + \xi_n\Omega^{(n)}$$

of a vector Φ on this space can be found by solving the n linear equations

$$(12.6) \qquad \sum_{\nu=1}^{n} (\Omega^{(\lambda)}, \Omega^{(\nu)}) \xi_\nu = (\Omega^{(\lambda)}, \Phi), \qquad \lambda = 1, \ldots, n.$$

That these equations have a solution follows from the fact that every linear relation that holds between the left members identically in $\xi_1, \ldots, \xi_n$ also holds between the right members.

Instead of verifying this fact -- which could easily be done -- we shall derive the statement from the fact that the vectors $\Omega^{(1)}, \ldots, \Omega^{(n)}$, assumed to be linearly independent can be chosen to be perpendicular to each other.

If these vectors are mutually orthogonal and normal, $||\Omega^{(\nu)}|| = 1$, the projector into the space $\mathfrak{P}_n$ spanned by them is simply given by

$$(12.7) \qquad P_n \Phi = \sum_{\nu=1}^{n} (\Omega^{(\nu)}, \Phi) \Omega^{(\nu)}.$$

Suppose the statement holds for $(n-1)$; then we may assume $\Omega^{(1)}, \ldots, \Omega^{(n-1)}$ to be mutually orthogonal and normal. Replacing $\Omega^{(n)}$ by $\Omega^{(n)} - P_{n-1}\Omega^{(n)}$ the statement follows for n.

13. Remarks about the Role of Self-Adjoint Operators in Physics

In Section 5 at the end of Chapter I, we have described three simple linear differential equations of physics in order to illustrate the use one can make of the spectral representation of the operator involved. To explain the role of self-adjointness in problems of physics let us consider more specifically the "wave equation"

$$\ddot{\Phi} + A\Phi = 0$$

for a vector $\Phi = \Phi(t)$, and show that the self-adjointness of the operator A implies the validity of the law of conservation of energy. It is mainly for this reason that the operators of physics are self-adjoint.

The (real) vector Φ here may stand for the displacement of a single mass particle, or of a system of k particles, from an undisplaced position: $\Phi = \{\phi_1,\ldots,\phi_k\}$. We may assume that these particles are connected by elastic springs or other mechanisms which try to bring them back to their original positions. Instead, we also may let the vector Φ stand for a function $\phi(x)$ which assigns to each point in a continuous elastic medium the displacement of the particle which originally was located at the point x.

In either case, the vector $-A\Phi$ stands for the forces per unit mass exerted on the displaced particles by the "restoring" mechanism.

In the linear space of displacement vectors let us introduce an inner product by taking as unit form the expression

$$(\Phi,\Phi) = \sum_{\kappa=1}^{k} \phi_\kappa^2$$

or, in case of a continuous medium $\mathscr{R}$, the expression

$$(\Phi,\Phi) = \int_{\mathscr{R}} \phi^2(x)\,dx,$$

where dx stands for a 3,2 or 1-dimensional differential according to the dimension of $\mathscr{R}$. Further, let us introduce a "mass operator" M, given by

$$M\Phi = \{m_1\phi_1,\ldots,m_\kappa\phi_\kappa\}$$

where $m_1,\ldots,m_\kappa$ are the masses of the k particles, or by

$$M\Phi = \mu(x)\phi(x)$$

where $\mu(x)$ is the mass density, i.e., the mass per unit volume, area, or length at the point x. Then the vector $-MA\Phi$ gives the restoring forces or force densities acting on the displaced particles.

Since this force vector $-MA\Phi$ depends linearly on the displacement Φ, the potential energy U stored in the restoring mechanism in the position on Φ can simply be expressed in the form

$$U = \frac{1}{2}(\Phi, MA\Phi),$$

with the aid of the inner product.

Suppose now the operator MA is formally self-adjoint, (i.e, symmetric, since the present vector space is real):

$$(\Phi', MA\Phi) = (MA\Phi', \Phi) .$$

Then the time-rate of change of the potential energy is

$$\dot{U} = \frac{1}{2}(\dot{\Phi}, MA\Phi) + \frac{1}{2}(MA\Phi, \dot{\Phi}).$$

Using the symmetry of the operator M we find the time rate of change of the total energy

$$E = \frac{1}{2}(\dot{\Phi}, M\dot{\Phi}) + \frac{1}{2}(\Phi, MA\Phi)$$

to be

$$\dot{E} = \frac{1}{2}(\dot{\Phi}, M\ddot{\Phi}+MA\Phi) + \frac{1}{2}(M\ddot{\Phi}+MA\Phi, \dot{\Phi}),$$

and hence zero by virtue of the differential equation. Thus, we see that the symmetry of the operator MA implies the constancy of the total energy E.

Also, for the solution $\Phi(t)$ of the Schrödinger equation

$$\dot{\Phi} + iA\Phi = 0$$

an important constancy relation obtains provided the operator A is formally self-adjoint, i.e., Hermitean, since we now assume the space to be complex. In fact, under this assumption we have

$$\frac{d}{dt}(\Phi,\Phi) = (\Phi,\dot{\Phi}) + (\dot{\Phi},\Phi) = (\Phi,iA\Phi) + (iA\Phi,\Phi)$$

$$= i\{(\Phi,A\Phi) - (A\Phi,\Phi)\} = 0,$$

so that

$$(\Phi,\Phi) = \text{constant}$$

for a solution of the Schrödinger equation. This result is vital since it is one of the stipulations of quantum theory that the vector Φ representing a state should have the norm 1. The Schrödinger equation should be consistent with this stipulation; if the norm $||\Phi||$ of its solution equals 1 initially, it should equal 1 at all times. That is the case if the operator A, the Hamiltonian energy operator, is Hermitean.

This fact is very important, but it is not the only important feature of quantum theory in which self-adjointness plays a vital role. We should like to discuss such other features.

When a quantity associated with a physical object is measured the outcome will depend on the "state" in which this object is. The states are identified with the elements of a vector space in which an inner product is defined. Properly speaking, the states are supposed to be vectors Φ with the norm 1; if the norm is not 1 one speaks of a "state vector" belonging to the state $\Phi/||\Phi||$. Any quantity, or "observable", is associated with a self-adjoint linear operator acting on the state vectors Φ. We simply use the same letter, A say, to denote the observable and the operator. The outcome of the measurement of such a quantity is not completely determined on the basis of the laws of physics as formulated by the quantum theory; but, the expectation value of this outcome is determined. It is given by the expression

$$(\Phi, A\Phi).$$

Since the operator A is assumed to be Hermitean, this expected value is a real number. This reality is one of the required features of quantum theory.

Another feature required of this theory is that to every -- say piecewise continuous -- function $f(\alpha)$ there should be assigned an observable $f(A)$ in such a way that measurement of this observable should give the value $f(\alpha)$ if measurement of A gave the value α. Such an assignment will be given by a "functional calculus" in the sense described in Section 3. Such a calculus is indeed possible if the operator A is self-adjoint in the strict sense; see Chapters IV and VI.

Consider, in particular, the characteristic function $\eta_{\Delta\alpha}(\alpha)$ of the interval $\Delta\alpha$; that is, the function which equals 1 for α inside $\Delta\alpha$, and 0 outside. The "spectral projector" $P_{\Delta\alpha} = \eta_{\Delta\alpha}(A)$ may also be regarded as an observable which can assume only the values 0 and 1. Measuring this value means to find out whether or not measurement of the observable A would yield a value inside or outside the interval $\Delta\alpha$. The expectation value $(\Phi, P_{\Delta\alpha}\Phi)$ of the observable $P_{\Delta\alpha}$ is then nothing but the probability that A would be found to have a value in $\Delta\alpha$. Thus, this probability is given by the expression $||P_{\Delta\alpha}\Phi||^2 = (\Phi, P_{\Delta\alpha}\Phi)$. The formulation of this basic contention involves the notion of spectral projector. It is clear that this formulation would not have been possible had it not been required that the operators which correspond to observables possess a spectral resolution.

Suppose the interval $\Delta\alpha$ is just a point α, and suppose an observation of the observable $P_{\Delta\alpha} = P_\alpha$ is to be made. If, before measurement, the object was in an eigenstate $\Phi = P_\alpha\Phi$ with the eigen-value α, the probability of finding the value α for A evidently equals 1. In such a case then, measurement of the observable A will yield with certainty the value α. This certainty of finding a definite value could never be attained for the improper eigen-values in a continuous spectrum.

CHAPTER III

HILBERT SPACE

14. Completeness

The statements about spectral representations which we have made up to now were not quite satisfactory insofar as the linear spaces in which the operators were supposed to act were specified in a rather arbitrary manner. For example, when we required continuous differentiability of the functions for which the Fourier series expansion was formulated we knew that this requirement was stronger than necessary; the situation was similar with the other spectral representations considered. The class of functions admitted could have been enlarged without modifying the formulation of the problem. These classes could have been taken still larger if one had adopted a generalized notion of convergence of a series or an integral. It is an important fact that this process of enlargement of the linear spaces has a definite end. There is a definite largest linear space in all spectral problems in which the spectral representation of an operator is possible. The property characterizing such space is the "completeness", which in turn involves the notions of "convergence" and "limit".

Let $\mathfrak{R}$ be a linear space of elements Φ provided with a norm $||\Phi||$ satisfying all the requirements postulated in Section 6. Then the notion of convergence is defined as follows:

A sequence of vectors Φ^{ν} converges to a vector Φ if

$$||\Phi^{\nu} - \Phi|| \to 0 \quad \text{as} \quad \nu \to \infty.$$

We note that the convergence of a sequence of vectors Φ^{ν} to a limit vector Φ^{ν} implies that their norms converge to that of the limit:

$$(14.2) \qquad ||\Phi^{(\nu)}|| \to ||\Phi|| \quad \text{as} \quad \nu \to \infty.$$

This follows immediately from relation (14.1) if one uses the second triangle inequality (6.5) in the form

$$\Big| \, ||\Phi^{(\nu)}|| - ||\Phi|| \, \Big| \leq ||\Phi^{(\nu)} - \Phi||.$$

In a space of finite dimension convergence in the sense here defined, means just convergence of each component, but in spaces of infinite dimension that is not so.

In the space of continuous functions $\phi(s)$ defined in the interval $\mathscr{I}$: $s_1 \leq s \leq s_2$, convergence with respect to the maximum norm $||\Phi|| = \max_{\mathscr{I}} |\phi(s)|$, is nothing but "uniform convergence", while convergence with respect to the norm

$$||\Phi|| = \left[\int_{\mathscr{I}} |\phi(s)|^2 ds \right]^{1/2},$$

is the same as convergence "in the mean":

$$\int_{\mathscr{I}} |\phi^{(\nu)}(s) - \phi(s)|^2 ds \to 0 \quad \text{as} \quad \nu \to \infty.$$

Certainly mean convergence is weaker than "uniform" convergence since, clearly, there are sequences of functions which converge in the mean but not uniformly.

Convergence as introduced presupposes a given limit vector. A notion of convergence can also be introduced without reference to a limit vector. We say, a sequence of vectors $\Phi^{(\nu)}$ converges "in itself" or it is a "Cauchy sequence", if

$$(14.3) \qquad ||\Phi^{(\nu)} - \Phi^{(\mu)}|| \to 0 \quad \text{as} \quad \nu, \mu \to \infty,$$

i.e., if to every $\varepsilon > 0$ there is a ν_ε such that

$$(14.3)' \qquad ||\Phi^{(\nu)} - \Phi^{(\mu)}|| \leq \varepsilon \quad \text{whenever} \quad \mu \geq \nu \geq \nu_\varepsilon.$$

From the second triangle inequality we conclude that the norms of the vectors of a Cauchy sequence themselves form a Cauchy sequence:

$$(14.4) \qquad ||\Phi^{(\nu)}|| - ||\Phi^{(\mu)}|| \to 0 \quad \text{as} \quad \nu, \mu \to \infty.$$

It is a consequence of this fact that the norms of the vectors of a Cauchy sequence are bounded.

It is clear that every sequence which converges to a limit vector also converges in itself. One need only determine ν_ε such that $||\Phi^{(\nu)} - \Phi|| \leq \varepsilon/2$ when $\nu \geq \nu_\varepsilon$; then $||\Phi^{(\nu)} - \Phi^{(\mu)}|| = ||\Phi^{(\nu)} - \Phi + \Phi - \Phi^{(\mu)}|| \leq \varepsilon$ for $\mu \geq \nu \geq \nu_\varepsilon$ by the triangle inequality (6.4).

The converse need not be true, however. There may exist in some spaces Cauchy sequences without limit vectors.

For example, it is easy to construct a sequence of continuous functions $\phi^{(\nu)}$ with

$$\int_{\mathscr{D}} |\phi^{(\nu)}(s) - \phi^{(\mu)}(s)|^2 ds \to 0, \quad \text{as} \quad \nu, \mu \to \infty.$$

to which there is no continuous limit function. We may simply take the sequence of functions

$$\phi^{(\nu)}(s) = 0, \ s \leq 0, \qquad = \nu s, \ 0 \leq s \leq 1/\nu, \qquad = 1, \quad 1/\nu \leq s \leq 1$$

defined in the interval $\mathscr{I}$: $-1 \leq s \leq 1$. Obviously, this sequence converges in itself with respect to the square integral norm. If it had a continuous limit function $\phi(s)$, clearly, one would have

$$\int_{-1}^{0} |\phi(s)|^2 ds = 0 \quad \text{and} \quad \int_{0}^{1} |\phi(s) - 1|^2 ds = 0;$$

or

$$\phi(s) = 0, \ s < 0, \ \phi(s) = 1, \ s > 0,$$

and this function is not continuous.

Of course, one could extend the space of functions by admitting piecewise continuous functions. Then the sequence just considered would have a limit function. However, it would again be possible to construct a Cauchy sequence of piecewise continuous functions without a piecewise continuous limit function. We shall see later on, in Section 15, that the function space can nevertheless be so extended that every Cauchy sequence has a limit.

A space in which every Cauchy sequence of vectors has a limit vector is called "complete".

A complete normed space is called a "Banach space".

For example, the space of continuous functions $\phi(s)$ in a closed interval $\mathscr{I}$ is complete with respect to the maximum norm $||\phi|| = \max_{\mathscr{I}} |\phi(s)|$; it hence is a Banach space. For, every sequence of functions which is a Cauchy sequence with respect to this norm, i.e., which is a uniform Cauchy sequence, has a limit function which again is continuous.

With the aid of the notion of completeness we can formulate the notion of "Hilbert space": <u>A Hilbert space is a complete inner product space</u>. Here completeness is supposed to refer to the norm

$||\Phi|| = [(\Phi,\Phi)]^{1/2}$ associated with the inner product.

According to this definition a Hilbert space may be of finite, countable, or non-countable dimension. Originally, the term Hilbert space was reserved by von Neumann for the space of countable dimension. The terminology here adopted is convenient, and rather commonly used now.

The case considered by Hilbert himself was a special case of countable dimension, viz. the space of sequences $\Phi = \{\xi_1, \xi_2, \ldots\}$ for which the series $\sum_\kappa |\xi_\kappa|^2$ converges to a finite limit. It is not difficult to prove that this space is linear, and that the expression

$$(\Phi',\Phi) = \sum_\kappa \overline{\xi'_\kappa}\xi_\kappa$$

always converges for vectors Φ, Φ' in this space and may serve as an inner product so that the norm becomes

$$||\Phi|| = \left[\sum_\kappa |\xi_\kappa|^2\right]^{1/2}.$$

We shall not give a proof of these statements here, since these statements will result as a special case of more general statements to be proved in Section 15.

What about function spaces? Since the space of continuous, or even the space of piecewise continuous, functions $\phi(s)$ defined in our interval $\mathscr{I}$ is not complete (with respect to the inner product norm) we may wonder whether or not this space can be enlarged to a complete one. This is indeed possible. Such a complete extension is obtained in the manifold of all functions $\phi(s)$ in $\mathscr{I}$ whose absolute square $|\phi(s)|^2$ is integrable in the sense of Lebesgue. The completeness of the resulting function space $\mathscr{L}_2$ is expressed by the statement that to every sequence $\phi(s)$ in $\mathscr{L}_2$ for which

$$\int_{\mathscr{I}} |\phi^\nu(s) - \phi^\mu(s)|^2 ds \to 0, \quad \nu, \mu \to \infty,$$

there is a function $\phi(s)$ in the space $\mathscr{L}_2$ such that

$$\int_{\mathscr{I}} |\phi^\nu(s) - \phi(s)|^2 ds \to 0, \quad \nu \to \infty.$$

This statement is a part of the celebrated theorem of Fischer and F. Riesz.

We could rely on this statement if we wished; but it is not necessary to do so. It is possible to attain the completion of function spaces directly, without invoking the theory of Lebesgue integration. This will be shown in the following section.

15. First Extension Theorem. Ideal Functions

In this section we shall show that every inner product space $\mathfrak{H}'$ can be extended to a complete one, a Hilbert space $\mathfrak{H}$, in which it is dense. An inner product space will, therefore, also be called an "Pre-Hilbert Space".

We recall that the subset $\mathfrak{B}'$ of a space $\mathfrak{B}$ was said to be dense in it if to every vector in $\mathfrak{B}$ there is an arbitrarily close vector in $\mathfrak{B}'$; in other words if every vector in $\mathfrak{B}$ is the limit of a sequence of vectors in $\mathfrak{B}'$; see Section 6.

First Extension Theorem. Let $\mathfrak{H}'$ be an inner product space. Then there exists a Hilbert space $\mathfrak{H}$ which contains $\mathfrak{H}'$ densely in such a way that the inner product defined in $\mathfrak{H}$ agrees with that originally defined in $\mathfrak{H}'$.

To establish this extension, let ϕ^ν be a Cauchy sequence of vectors in $\mathfrak{H}'$. To such a Cauchy sequence we assign an "ideal element", or "ideal vector" denoted by Φ. We assign the same ideal element to two Cauchy sequences $\{\Phi_1^\nu\}$ and $\{\Phi^\nu\}$ provided $||\Phi_1^\nu - \Phi^\nu|| \to 0$ as $\nu \to \infty$; we call two such sequences "equivalent". In other words, each ideal vector corresponds to a class of equivalent Cauchy sequences. Every vector in the space $\mathfrak{H}'$ itself may be regarded as an ideal vector; for every such vector Φ' generates the Cauchy sequence $\Phi',\Phi',\Phi',\ldots$ and we simply identify the corresponding ideal vector with Φ'. Having done so we may say that the set of ideal vectors contains $\mathfrak{H}'$ as a subspace.

Note that the completion process described is the precise analogue to one of the processes by which the set of rational numbers can be extended to the set of rational and irrational real numbers.

Of course, we must show that the set $\mathfrak{H}$ of ideal vectors forms a linear space; furthermore, we must define an inner product in it and show that it has the desired properties.

Let Φ and $\tilde{\Phi}$ be two (ideal) vectors in the extension $\mathfrak{H}$ given by two Cauchy sequences $\{\Phi^\nu\}$ and $\{\tilde{\Phi}^\nu\}$ taken from $\mathfrak{H}'$.

Let $c,\tilde{c}$ be two complex numbers; then the sequence $\{c\Phi^\nu + \tilde{c}\tilde{\Phi}^\nu\}$ is also a Cauchy sequence, as follows from the triangle inequality. We should like to denote the associated ideal element by $\{c\Phi + \tilde{c}\tilde{\Phi}\}$.

Furthermore, the sequence of numbers $(\tilde{\Phi}^\nu,\Phi^\nu)$ is a Cauchy sequence since by virtue of the Schwarz inequality the estimate

$$|(\tilde{\Phi}^\nu,\Phi^\nu) - (\tilde{\Phi}^\mu,\Phi^\mu)| = |(\tilde{\Phi}^\nu,\Phi^\nu - \Phi^\mu) + (\tilde{\Phi}^\nu - \tilde{\Phi}^\mu,\Phi^\mu)|$$

$$\leq ||\tilde{\Phi}^\nu||\ ||\Phi^\nu - \Phi^\mu|| + ||\tilde{\Phi}^\nu - \tilde{\Phi}^\mu||\ ||\Phi^\mu||$$

holds and since $||\tilde{\Phi}^\nu||$ and $||\Phi^\mu||$ are bounded as shown above. Consequently, the numbers $(\tilde{\Phi}^\nu,\Phi^\nu)$ approach a limit which we should like to denote by $(\tilde{\Phi},\Phi)$.

We must make sure that the assignments of $c\Phi + \tilde{c}\tilde{\Phi}$ and of $(\tilde{\Phi},\Phi)$ to Φ and $\tilde{\Phi}$ are independent of the choice of the defining Cauchy sequence $\{\Phi^\nu\}$ and $\{\tilde{\Phi}^\nu\}$. To this end we make the following obvious but useful remarks.

1) Every subsequence of a Cauchy sequence is again a Cauchy sequence, equivalent to the full sequence.

2) The mixture of two equivalent Cauchy sequences, constructed by taking, alternatingly, one term from the first and one from the second, is again a Cauchy sequence equivalent to each of the components.

Now we first observe that the limit of the numbers $(\tilde{\Phi}^\nu,\Phi^\nu)$ is unchanged if we restrict Φ^ν and $\tilde{\Phi}^\nu$ to subsequences. Next, we consider the mixture of the sequence $\{\Phi^\nu\}$ and an equivalent sequence $\{\Phi_1^\nu\}$. Since the mixtures are Cauchy sequences the inner products for them have limits; since the mixtures are equivalent with the components, the limit of the inner products for the mixture is the same as that found with the original and with the new components. Hence the limit of the inner product is independent of the choice of the defining Cauchy sequence.

The same argument, of course, applies to the linear combination.

Having assigned a linear combination $(c\Phi+\tilde{c}\tilde{\Phi})$ and an inner product to the ideal elements we should verify that these assignments have the required properties. This could be done easily. We shall carry out such a verification only for one of the properties, viz. the property (7.4) that $(\Phi,\Phi) = 0$ implies $\Phi = 0$.

To this end we note that $(\Phi,\Phi) = 0$ means $\Phi^\nu \to 0$ as $\nu \to \infty$ where $\{\Phi^\nu\}$ is the sequence defining Φ. Let $\{0^\nu\}$ be the Cauchy sequence consisting of the zero vectors, $0^\nu = 0$, then we have $||\Phi^\nu - 0^\nu|| = ||\Phi^\nu|| \to 0$. Hence $\{\Phi^\nu\}$ and $\{0^\nu\}$ give the same ideal element; but $\{0^\nu\} = \{0,0,0 \ldots\}$ was identified with 0.

We have now come to the conclusion that the extension $\mathfrak{H}$ of $\mathfrak{H}'$ is a linear inner product space. We still must show that this space is complete.

Let $\{\Phi^\nu\}$ be a Cauchy sequence of ideal elements in $\mathfrak{H}$. Then let $\tilde{\Phi}^\nu$ be an element in $\mathfrak{H}'$ such that $||\tilde{\Phi}^\nu - \Phi^\nu|| \leq 1/\nu$. From the triangle inequality we have

$$||\tilde{\Phi}^\nu - \tilde{\Phi}^\mu|| \leq ||\Phi^\nu - \Phi^\mu|| + 2/\nu \to 0$$

so that $\{\tilde{\Phi}^\nu\}$ is seen to be a Cauchy sequence, defining an element Φ. One easily verifies $||\Phi^\nu - \Phi|| \to 0$.

Finally, we note that the space $\mathfrak{H}'$ is dense in the extension $\mathfrak{H}$, as is evident from the construction of $\mathfrak{H}$.

Thus we have established the first extension theorem. We add the obvious

<u>Remark</u>: The completion of a space is the same as that of each dense subspace of it. This remark allows one considerable leeway in the choice of the subspaces to attain a desired complete space.

As a <u>first application</u> of the first extension theorem we let $\mathfrak{H}'$ be <u>the space of all sequences</u>

$$\Phi = \{\xi_1, \xi_2, \ldots, 0, 0, \ldots\}$$

of a finite number of components, the number of which is not restricted, and take the (finite) series

$$(\Phi,\Phi) = \sum_{\kappa=1}^{\infty} |\xi_\kappa|^2 \tag{15.1}$$

as unit form. We maintain that <u>the ideal elements which form the</u>

completion $\mathfrak{H}$ of $\mathfrak{H}'$ can be realized as (infinite or finite) sequences $\{\xi_\kappa\}$ of numbers ξ_κ for which $\sum_{\kappa=1}^{\infty} |\xi_\kappa|^2$ is finite. In other words, we maintain that this completion is the original "special" Hilbert space, mentioned in Section 14.

It is clear that every sequence $\{\xi_1,0,0,\ldots\}$, $\{\xi_1,\xi_2,0,0,\ldots\}$, $\{\xi_1,\xi_2,\xi_3,0,\ldots\}$ forms a Cauchy sequence provided $\sum_\kappa |\xi_\kappa|^2 < \infty$; all that we have to show is that every Cauchy sequence of vectors in $\mathfrak{H}'$ is equivalent to such a special one. That is to say we should prove that to every Cauchy sequence $\{\Phi^{(\alpha)}\}$ of vectors in $\mathfrak{H}'$ a sequence $\{\Phi_k\}$ of vectors $\Phi_k = \{\xi_1,\ldots,\xi_k,0,0,\ldots\}$ can be assigned such that

$$||\Phi^{(\alpha)} - \Phi_{k(\alpha)}|| \to 0$$

for $\alpha \to \infty$ and appropriate $k(\alpha) \to \infty$.

To this end we note that the relation $||\Phi^{(\alpha)} - \Phi^{(\beta)}|| \to 0$ implies that to every $\varepsilon > 0$ there is an $\alpha(\varepsilon)$ such that

$$||\Phi^{(\alpha)} - \Phi^{(\beta)}|| \leq \varepsilon \quad \text{for} \quad \beta \geq \alpha \geq \alpha(\varepsilon).$$

Set $\Phi^{(\alpha)} = \{\xi_1^{(\alpha)},\xi_2^{(\alpha)},\ldots,\xi_\kappa^{(\alpha)},\ldots,0,0,\ldots\}$ and denote by $\Phi_k^{(\alpha)} = \{\xi_1^{(\alpha)},\ldots,\xi_k^{(\alpha)},0,0,\ldots\}$ the vector obtained from $\Phi^{(\alpha)}$ by replacing all components $\xi_\kappa^{(\alpha)}$ by 0 for $\kappa > k$. Then, clearly, $||\Phi_k^{(\alpha)} - \Phi_k^{(\beta)}|| \leq \varepsilon$ and hence, as $\alpha \to \infty$, the vectors $\Phi_k^{(\alpha)}$ converge to a limit vector $\Phi_k = \{\xi_1,\ldots,\xi_k,0,\ldots\}$. Evidently, the components ξ_κ of Φ_k are independent of k for $\kappa \leq k$. Letting β tend to ∞ in the above inequality we find $||\Phi_k^{(\alpha)} - \Phi_{k(\alpha)}|| \leq \varepsilon$. We now choose $k = k(a)$ so large that the components ξ_κ of $\Phi^{(\alpha)}$ vanish for $\kappa \geq k(\alpha)$. Then $\Phi_{k(\alpha)}^{(\alpha)} = \Phi^{(\alpha)}$ and hence

$$||\Phi^{(\alpha)} - \Phi_{k(\alpha)}|| \leq \varepsilon \quad \text{for} \quad \alpha \geq \alpha(\varepsilon).$$

Thus we have attained our goal.

It is thus clear that the space of infinite sequences $\Phi = \{\xi_1, \xi_2, \ldots\}$ with

$$(15.2) \qquad (\Phi,\Phi) = \sum_{\kappa=1}^{\infty} |\xi_\kappa|^2$$

is the extension $\mathfrak{H}$ of the space $\mathfrak{H}'$ of finite sequences and hence a Hilbert space. As said before this was the original space of infinitely many variables investigated by Hilbert.

A very important application of the first extension theorem is the completion of function spaces. For example, we may complete the space of piecewise continuous functions $\phi(s)$ defined in a finite interval $\mathscr{R}$ of the s-axis, and carrying the unit form

$$(15.3) \qquad (\phi,\phi) = \int_{\mathscr{R}} |\phi(s)|^2 dr(s),$$

involving a measure function $r(s)$. The completion of this space consists in adjoining "ideal elements" in the sense described above.

In case the domain $\mathscr{R}$ extends to infinity we may begin our extension with the space $\dot{\mathfrak{C}}'$ of piecewise continuous functions with bounded support. We then may carry out our extension in two stages and first extend the space $\dot{\mathfrak{C}}'$ to the space $\mathfrak{C}' \cap \mathfrak{H}$ of those piecewise continuous functions $\phi(s)$ for which

$$\int_{\mathscr{R}} |\phi(s)|^2 dr(s) < \infty.$$

Clearly, any such function may be identified with the ideal element which is given by the Cauchy sequence formed by the functions

$\phi^{\nu}(s) = \phi(s)$ for $|s| < s_{\nu}$, $= 0$ for $|s| \geq s_{\nu}$, $s_{\nu} \to \infty$, belonging to $\dot{\mathfrak{C}}'$. The space $\mathfrak{C}' \cap \mathfrak{H}$ is therefore an inner product space which now may be completed to the space $\mathfrak{H}$.

The "ideal elements" entering the completion of a function space will be called "ideal functions". As a matter of fact, we shall refer to these elements simply as "functions", and use the notion $\phi(\cdot)$ or even $\phi(s)$ for them, although most of these "functions" do not assign definite values to the values of s in interval $\mathscr{R}$. It is true, some of these "functions" may be materialized by functions that have definite values everywhere in $\mathscr{R}$. In fact, by virtue of the Lebesgue theory, realization of ideal functions is always possible. But, we do not make use of such realizations; at least not for most of those parts of spectral theory we shall deal with. In case we need such a realization, we shall make special provisions.

We say that an ideal function ϕ is zero, $\phi = 0$, if

$$\int_{\mathscr{R}} |\phi(s)|^2 dr(s) = 0.$$

This does not mean that the values of any proper function $\phi(s)$ which affords a realization of the ideal function ϕ vanish for every value of s. In fact, as we had seen earlier, in Chapter II, Section 8, the values of a piecewise continuous function outside of the carrier of $r(s)$ do not contribute to the integral (ϕ,ϕ), so that $\phi = 0$ if ϕ vanishes in the carrier of r. Also, $\phi = 0$ if $\phi(s) = 0$ except at some values of s where $r(s)$ is continuous. In other words, the situation that $\phi = 0$ does not imply $\phi(s) \equiv 0$ arose already with proper functions.

While we do not want to make use of the possibility to ascribe values to the ideal functions at each point s, we do want to be able to say <u>when an ideal function vanishes identically in a subinterval</u> $\mathscr{I}$ of the interval $\mathscr{R}$.

Let $\eta(s)$ be the characteristic function of the interval $\mathscr{I}$, i.e., let $\eta(s) = 1$ for s in $\mathscr{I}$, $= 0$ otherwise. Then the sequence of piecewise continuous functions $\eta(s)\phi^{\nu}(s)$ forms a Cauchy sequence if the sequence $\{\phi^{\nu}(s)\}$ does. The limit function of this sequence will be denoted by $\eta(s)\phi(s)$ when $\phi(s)$ is the limit function of $\{\phi^{\nu}(s)\}$. Evidently, multiplication by $\eta(s)$ induces a projector, $P_{\mathscr{I}}$, which thus is defined in the whole complete space of ideal functions.

Having defined multiplication by $\eta(s)$, we may say that the ideal function vanishes in the interval $\mathscr{I}$ if $\eta\phi = 0$; i.e., if $P_{\mathscr{I}}\phi = 0$.

Suppose the interval $\mathscr{I}$ consists just of the point $s = s_o$ which is not a jump point of the measure function $r(s)$.

Then, we maintain, $P_{\mathscr{I}} = P_{s_o} = 0$.

Clearly, for any piecewise continuous function $\phi^{\nu}(s)$ approximating an ideal function ϕ we have

$$\int_{\mathscr{R}} |\eta(s)\phi^{\nu}(s)|^2 dr(s) = \int_{\mathscr{I}} |\phi^{\nu}(s)|^2 dr(s) = 0$$

so that $P_{\mathscr{I}}\phi^{\nu} = 0$ and hence $P_{\mathscr{I}}\phi = 0$.

In this context the following question arises. Suppose the ideal function ϕ is identically zero in every interval not containing the point s_o, is it identically zero? We want to show that this is indeed the case if s_o is not a jump point of $r(s)$. In other words, we want to prove the

<u>Remark</u>: Suppose the ideal function ϕ vanishes in every interval not containing a point s_o assumed not to be a jump point of $r(s)$. Then $\phi = 0$.

Proof. Let $\phi^\nu(s)$ be a sequence of piecewise continuous functions which approximates $\phi(s)$, so that to every $\varepsilon \geq 0$ there is a ν such that

$$||\phi^\nu - \phi||^2 = \int |\phi^\nu(s) - \phi(s)|^2 dr(s) < \varepsilon^2.$$

Furthermore, since $\phi^\nu(s)$ is piecewise continuous and s_o is not a jump point of $r(s)$, a number $\sigma > 0$ can be chosen such that

$$\int_{|s-s_o|\leq\sigma} |\phi^\nu(s)|^2 dr(s) \leq \varepsilon^2.$$

By assumption we have

$$\int_{|s-s_o|>\sigma} |\phi(s)|^2 dr(s) = 0$$

and, therefore

$$\int_{|s-s_o|>\sigma} \overline{\phi(s)}\ \phi^\nu(s) dr(s) = 0.$$

Consequently, we have

$$||\phi||^2 = \int_{\mathscr{R}} \overline{\phi(s)}\ (\phi(s) - \phi^\nu(s)) dr(s) + \int_{|s-s_o|\leq\sigma} \bar{\phi}(s)\phi^\nu(s) dr(s)$$

$$\leq ||\phi||\varepsilon + ||\phi||\varepsilon$$

whence $||\phi|| \leq \varepsilon$. The statement of the remark then follows.

Ideal functions are particularly appropriate entities to employ in those branches of physics which involve linear operators, in particular in quantum theory. In their mathematical operations physicists have always been guided by the conviction that formal operations are justified somehow or other; to a large extent such formal operations can be justified by interpreting the functions involved as ideal ones.

To be sure, in the end one wants to work with proper functions which assign definite values to values of the independent variables; and in fact, the solutions of most concrete problems can in the end be shown to be given by proper functions. But, it is of greatest advantage, not to insist too early on showing that the ideal functions one works with are proper.

It should be mentioned that the "distributions" of L. Schwartz represent a class of "ideal functions" less restricted than the class of ideal functions considered here, inasmuch as they are not related to a norm; but rather to a particular seminorm. To be sure, there are questions in spectral theory in which distributions of certain kinds are the appropriate entities to employ; in fact, we shall employ distributions in later chapters. For most of our purposes, however, it is exactly the restriction to a Hilbert space which makes the ideal functions as we have introduced them an effective tool.

16. Fourier Transformation

The role of ideal functions may be illustrated in connection with the Fourier series expansion discussed in our second example in Section 1. There we had considered a space of vectors Φ given by functions $\phi(s)$ defined for $-\pi \leq s \leq \pi$ and also represented by sequences η_μ, $\mu = 0, \pm 1, \pm 2, \ldots$. The transformation (1.9) and its inverse (1.9)* connecting these two representations are valid if the functions $\phi(s)$ are sufficiently smooth, for example, if they have continuous derivatives $d\phi(s)/ds$ and are periodic $\phi(\pi) = \phi(-\pi)$. Furthermore, the identity

(16.1) $$\frac{1}{2\pi}\int_{-\pi}^{\pi} |\phi(s)|^2 ds = \sum_{\mu=-\infty}^{\infty} |\eta_\mu|^2$$

holds for such functions.

We now know that we can extend the space $\mathfrak{H}'$ of these functions $\phi(s)$ to a Hilbert space $\mathfrak{H}$, consisting of ideal functions $\phi(\cdot)$ which may be realized by more or less smooth proper functions.

The sequence of components $\{\eta_\mu\}$ corresponding to the functions in $\mathfrak{H}'$ form a subspace of the complete space of all sequences $\{\eta_\mu\}$ for which $\sum_\mu |\eta_\mu|^2$ is finite. Clearly, the complete space corresponds to a complete space of such sequences, which we shall call admitted for the moment. We maintain that all sequences with finite $\sum_\mu |\eta_\mu|^2$ are admitted, i.e. that $\mathfrak{H}$ actually corresponds to the space of all such sequences, that is to the special Hilbert space of sequences.

To prove this we need only take an arbitrary set of numbers η_ν such that $\eta_\nu = 0$ for $|\nu| > k$ and assign it the function

$$\Phi_k = \phi_k(s) = \sum_{|\nu|\leq k} \eta_\nu e^{i\nu s} .$$

These functions are in the original function space $\mathfrak{H}'$; hence it follows that all finite sequences are admitted. Since the space of all sequences is the closure of the space of the finite ones, our contention is proved.

Thus we see that the complete space of functions $\phi(s)$ corresponds to the complete space of all square summable sequences.

The statement expressing this correspondence is essentially that part of the Fischer-Riesz theorem which is concerned with Fourier series.

We should note that the inverse transformation (1.9)* is

meaningful as it stands even in case the function $\phi(s)$ entering in it is any ideal function of the complete spaces $\mathfrak{H}$; for, the integral involved in this formula may be regarded as the inner product of the two functions $\phi(s)$ and $e^{i\mu s}$, both being in $\mathfrak{H}$. On the other hand, the direct transformation formula (1.9) cannot be interpreted as an inner product; it may, however, be interpreted as the symbolic expression of the statement that the ideal function $\phi(\cdot)$ can be approximated in the sense of the norm by the function

$$\phi_k(s) = \sum_{|\mu| \leq k} \eta_\mu e^{i\mu s}$$

when η_μ is given by (1.9)*.

The situation is similar, but not quite as simple, in the third example, concerned with the Fourier integral transformation. Before discussing it, we like to formulate a general lemma, which describes the structure of the argument to be used.[†]

Lemma. Let $\mathfrak{H}$ and $\mathfrak{G}$ be two Hilbert spaces, $\mathfrak{H}'$ and $\mathfrak{G}'$ dense subspaces in them. Let T and S be linear transformations mapping respectively $\mathfrak{H}'$ into $\mathfrak{G}$, $\mathfrak{G}'$ into $\mathfrak{H}$, and having the following properties:

(16.2) $\quad (\Phi, S\Psi) = (T\Phi, \Psi)$ for Φ in $\mathfrak{H}'$, Ψ in $\mathfrak{G}'$,

$(16.3)_T$ $\quad ||T\Phi|| = ||\Phi||$ for Φ in $\mathfrak{H}'$,

$(16.3)_S$ $\quad ||S\Psi|| = ||\Psi||$ for Ψ in $\mathfrak{G}'$.

[†]This argument follows a suggestion by P. Rejto.

Then the transformations S and T can be extended as linear operators to $\mathfrak{G}$ and $\mathfrak{H}$ such that the same three relations hold for all Φ in $\mathfrak{H}$, Ψ in $\mathfrak{G}$. Furthermore, the relations

$$(16.4)_T \qquad ST\Phi = \Phi \quad \text{for} \quad \Phi \quad \text{in} \quad \mathfrak{H},$$

$$(16.4)_S \qquad TS\Psi = \Psi \quad \text{for} \quad \Psi \quad \text{in} \quad \mathfrak{G},$$

hold, which express the fact that S is the inverse of T and vice versa.

Since $\mathfrak{H}'$ is dense in $\mathfrak{H}$ we may find to every Φ in $\mathfrak{H}$ a sequence Φ^ν in $\mathfrak{H}'$ such that $||\Phi^\nu - \Phi|| \to 0$ as $\nu \to \infty$. From $(16.3)_T$ it then follows that the vectors $T\Phi^\nu$ converge to a vector in $\mathfrak{G}$ which we denote by $T\Phi$. Clearly, this vector is independent of the choice of the sequence Φ^ν; it then follows that the operator T now defined in $\mathfrak{H}$ is linear. Similarly, we may extend the operator S to all of $\mathfrak{G}$ so as to be linear there. Evidently, the relations (16.2) and (16.3) hold for the operators T and S with all vectors Φ in $\mathfrak{H}$, Ψ in $\mathfrak{G}$.

Using $(16.3)_S$ and (16.2) with $\Psi = T\Phi$ and further $(16.3)_T$ we derive the relation

$$\begin{aligned} ||ST\Phi-\Phi||^2 &= ||ST\Phi||^2 - (ST\Phi,\Phi) - (\Phi,ST\Phi) + ||\Phi||^2 \\ &= ||T\Phi||^2 - 2(T\Phi,T\Phi) + ||\Phi||^2 = ||\Phi||^2 - ||T\Phi||^2 = 0, \end{aligned}$$

which shows that indeed $(16.4)_T$ holds. Similarly $(16.4)_S$ follows.

These two relations, incidentally, imply that the range of the operators T and S acting in the complete spaces $\mathfrak{G}$ and $\mathfrak{H}$ are the complete spaces $\mathfrak{G}$ and $\mathfrak{H}$ respectively, since every Φ can be written in the form $\Phi = S(T\Phi)$ or every Ψ in the form

$\Psi = T(S\Psi)$.

The lemma can, of course, also be used for the Fourier series transformation. We may take as $\mathfrak{G}'$ the space of all sequences with only a finite number of non-vanishing components. As space $\mathfrak{H}'$ we may take the space of periodic functions with periodic continuous first derivatives. The transformation (1.9) and its inverse (1.9)* may be taken as S and T. The unit forms are the two sides of (16.1). Properties (16.2) and $(16.3)_S$ are then immediately verified and property $(16.3)_T$ may be taken from the theory of Fourier series, or derived by arguments similar to those employed below for the Fourier integral transformation. The extension to the complete spaces $\mathfrak{H}$ and $\mathfrak{G}$ is thus seen to be possible.

In case of the Fourier integral transformation, we may take as space $\mathfrak{H}'$ the space of all functions $\phi(s)$ with bounded support having continuous first derivatives. The space $\mathfrak{G}'$ of functions $\eta(s)$ is defined correspondingly. The unit forms will be taken as

$$(16.5) \qquad (\eta,\eta) = \int_{-\infty}^{\infty} |\eta(\mu)|^2 d\mu, \quad (\phi,\phi) = \frac{1}{2\pi}\int_{-\infty}^{\infty} |\phi(s)|^2 ds.$$

The transformations

$$(16.6) \qquad S\eta(s) = \int_{-\infty}^{\infty} e^{is\mu}\eta(\mu)\,d\mu, \quad T\phi(\mu) = \frac{1}{2\pi}\int_{-\infty}^{\infty} e^{-i\mu s}\phi(s)\,ds$$

then produce continuous functions of s and μ which decay at infinity at least of order $|s|^{-1}$ and $|\mu|^{-1}$ respectively. Clearly, these functions are respectively in $\mathfrak{H}$ and $\mathfrak{G}$.

Property (16.2) is immediately verified. To verify identities (16.3), which assume the form

$$(16.7) \qquad \frac{1}{2\pi}\int_{-\infty}^{\infty} |\phi(s)|^2 ds = \int_{-\infty}^{\infty} |\eta(\mu)|^2 d\mu,$$

we could rely on the theory of the Fourier transformation. Still, we shall give a brief derivation of them. In the relation

$$4\pi^2 \int_{-m}^{m} |T\phi(\mu)|^2 d\mu = \int_{-m}^{m} \int_{-\infty}^{\infty} \int_{-\infty}^{\infty} e^{i(s'-s)\mu} \phi(s)\phi(s')ds'dsd\mu$$

we may interchange the order of integration since the function $\phi(s)$ has bounded support. Assuming $\phi(s) = 0$ for $|s| \geq \sigma$ we obtain for the left member the expression

$$2 \int_{-\sigma}^{\sigma} \int_{-\sigma}^{\sigma} \frac{\sin m(s'-s)}{s'-s} \phi(s')\overline{\phi(s)}ds'ds$$

$$= 2 \int_{-\sigma}^{\sigma} \int_{-\sigma}^{\sigma} \sin m(s'-s) \frac{\phi(s') - \phi(s)}{s'-s} \phi(s)ds'ds$$

$$+ 2 \int_{-\sigma}^{\sigma} \left[\int_{-\sigma-s}^{\sigma-s} \frac{\sin ms''}{s''} ds'' \right] |\phi(s)|^2 ds.$$

Note that the integral over s'' tends to the value π as m tends to infinity for $-\sigma < s < \sigma$, since then $-\sigma - s < 0$ and $\sigma - s > 0$. The last term, therefore, tends to $2 \int_{-\sigma}^{\sigma} |\phi(s)|^2 ds$. The first term tends to zero as $m \to \infty$ by the Riemann-Lebesgue Lemma since the difference quotient $(\phi(s') - \phi(s))/(s'-s)$ is continuous by virtue of the assumed continuous differentiability of $\phi(s)$. Hence the desired relation

$$2\pi \int_{-\infty}^{\infty} |T\phi(\mu)|^2 d\mu = \int_{-\sigma}^{\sigma} |\phi(s)|^2 ds$$

ensues; i.e. we have established relation $(16.3)_T$ for Φ in $\mathfrak{H}'$. Relation $(16.3)_S$ for Ψ in $\mathfrak{G}$ is established at the same time.

It is possible to avoid use of the Riemann-Lebesgue Lemma in the following way. Instead of letting the spaces $\mathfrak{H}'$ and $\mathfrak{G}'$ consist of differentiable functions we simply let these spaces consist of all piecewise constant functions of bounded support. These functions are finite linear combinations of the unit step functions associated with intervals $\mathscr{I}$. In order to prove identity $(16.3)_S$ for both functions it is then evidently sufficient to establish the relation

$$\frac{1}{2\pi}\int_{-\infty}^{\infty} \overline{S\eta_1(s)}S\eta_2(s)\,ds = \int_{-\infty}^{\infty} \overline{\eta_1(\mu)}\eta_2(\mu)\,d\mu = \Delta_{12}$$

where Δ_{12} is the length of the intersection of the intervals $\mathscr{I}_1$ and $\mathscr{I}_2$ associated with $\eta_1(\mu)$ and $\eta_2(\mu)$. Now, since, $S\eta(s) = (is)^{-1}[e^{is\mu^+} - e^{-is\mu^-}]$ where $\mu^{\pm}$ are the end points of the interval $\mathscr{I}$, we need only prove the relation

$$\frac{1}{2\pi}\int_{-\infty}^{\infty}\left[e^{-is\mu_1^+} - e^{-is\mu_1^-}\right]\left[e^{is\mu_2^+} - e^{is\mu_2^-}\right] s^{-2}ds = \Delta_{12},$$

which is easily done by complex integration.

In both cases it is seen that the Lemma is applicable. Consequently, the Fourier transformation can be extended to the complete function spaces $\mathfrak{H}$ and $\mathfrak{G}$, in such a way that relations (16.2), (16.3) and (16.4) hold. Relations (16.6) may also be adopted in these complete spaces and regarded as a symbolic expression of the transformations S and T.

The result is related to the theorem of Plancherel, which describes the Fourier transformation in these complete spaces in a more specific way. In the present course we are satisfied with establishing basic relations (for the present as well as other cases) in subspaces of sufficiently smooth functions and their extension to complete spaces. A more specific description of the nature of such relations in the complete spaces will be given only if there are special reasons for doing so.

17. The Projection Theorem

The wider a space is, the easier should it be to find in it an entity with desired properties. Earlier, in Section 12, we discussed the operation of orthogonal projection of a vector into a subspace in an inner product space and asked whether or not there always is such a projection. We shall show that indeed there is always such projection if the subspace is "complete". Here then we shall be rewarded for our efforts in making spaces complete. The statement is embodied in the "Projection Theorem", the basic theorem of the geometry of the Hilbert space.

Projection Theorem. Every vector in an inner product space possesses a unique orthogonal projection on any complete subspace.

In general, the inner product space $\mathfrak{B}$ will itself be complete, i.e. a Hilbert space.

We recall that the projection $P\Phi$ of a vector Φ on a subspace $\mathfrak{F}$ is a vector in $\mathfrak{F}$ such that $\Phi - P\Phi$ is orthogonal to all vectors Ψ in $\mathfrak{F}$. We maintain that the distance of the vector Φ from any vector Ψ' in $\mathfrak{F}$ other than $P\Phi$ is greater than its distance from $P\Phi$; for,

$$\begin{aligned} ||\Phi - \Psi'||^2 &= ||(\Phi - P\Phi) + (P\Phi - \Psi')||^2 \\ &= ||\Phi - P\Phi||^2 + ||P\Phi - \Psi'||^2, \end{aligned}$$

since $\Phi - P\Phi$ is orthogonal to all vectors in $\mathfrak{F}$, and hence, in particular to $P\Phi - \Psi'$. This minimum property of the projection is the starting point for the proof of the projection theorem.

Consider the set of numbers $||\Phi - \Psi'||$ where Ψ' runs all over $\mathfrak{F}$. Certainly, this set of numbers has a greatest lower bound, d, and there is a sequence of vectors Ψ^ν in $\mathfrak{F}$ for which $||\Phi - \Psi^\nu||$ approaches this lower bound if $\nu \to \infty$,

(A) $$||\Phi - \Psi^\nu|| \geq d$$

and

(B) $$||\Phi - \Psi^\nu|| \to d \quad \text{as} \quad \nu \to \infty.$$

We call Ψ^ν a "minimizing" sequence. We want to prove that this sequence has a limit and that this limit is the desired projection.

To this end we use the following identity which holds for any triple of vectors Φ, Ψ', Ψ''

$$\frac{1}{2}||\Phi - \Psi'||^2 + \frac{1}{2}||\Phi - \Psi''||^2 = ||\Phi - \frac{1}{2}(\Psi' + \Psi'')||^2 + ||\frac{1}{2}(\Psi' - \Psi'')||^2,$$

and is immediately verified by working out the squares formally.

This identity may be related to the fact that the norm $||\Phi - \Psi||$ is a convex function of Ψ. We also mention incidentally that this identity has a simple geometric interpretation: The sums of the squares of lengths of the two diagonals of a parallelogram is the sum of the squares of the lengths of the sides.

In using this identity we note that $\Psi' + \Psi''$ is a vector in $\mathfrak{F}$ so that $||\Phi - \frac{1}{2}(\Psi' + \Psi'')|| > d$ by A, and hence

$$||\Psi' - \Psi''|| \leq 2||\Phi - \Psi'|| + 2||\Phi - \Psi''|| - 4d.$$

Taking $\Psi' = \Psi^{\nu}$ and $\Psi'' = \Psi^{\mu}$ and letting ν,μ tend to infinity we find, by B, the relation

$$||\Psi^{\nu} - \Psi^{\mu}|| \to 0, \quad \nu,\mu \to \infty.$$

This relation just says that $\{\Psi^{\nu}\}$ is a Cauchy sequence.

Now we make use of the assumed completeness of the space $\mathfrak{H}$ and conclude that there exists in $\mathfrak{F}$ a vector Ψ_o such that $||\Psi^{\nu} - \Psi_o|| \to 0$ as $\nu \to \infty$. This relation may also be written as $||(\Phi - \Psi^{\nu}) - (\Phi - \Psi_o)|| \to 0$, so that, using (14.2), we may conclude that the norm of $\Phi - \Psi^{\nu}$ tends to that of $\Phi - \Psi_o$:

$$||\Phi - \Psi^{\nu}|| \to ||\Phi - \Psi_o|| \quad \text{as} \quad \nu \to \infty.$$

Now since $||\Phi - \Psi^{\nu}||$ tends to d, by B, we have

(C) $$||\Phi - \Psi_o|| = d.$$

In other words, the greatest lower bound of $||\Phi - \Psi||$ is assumed; it is a minimum.

In a standard way we derive from this minimum property of Ψ_o that the "first variation" of the functional $||\Phi - \Psi||$ vanishes for $\Psi = \Psi_o$. By this we mean that the first derivative of the function $||\Phi - \Psi(t)||$ of t vanishes for $t = 0$ for every differentiable function $\Psi(t)$. Actually, it is sufficient to take $\Psi(t) = \Psi_o + t\Psi_1$ linear; then

$$||\Phi - \Psi(t)||^2 = ||\Phi - \Psi_o - t\Psi_1||^2 = ||\Phi - \Psi_o||^2$$

$$- 2t\mathrm{Re}(\Psi_1, \Phi - \Psi_o) + t^2||\Psi_1||^2.$$

Since by A and C this function attains a minimum for $t = 0$ we have $\mathrm{Re}(\Psi_1, \Phi - \Psi_o) = 0$ for all Ψ_1 in $\mathfrak{F}$. The same is true for the imaginary part of $(\Psi_1, \Phi - \Psi_o)$ since $i\Psi_1$ is also in $\mathfrak{F}$. Consequently we may conclude that

$$(\Psi, \Phi - \Psi_o) = 0 \quad \text{for all} \quad \Psi \quad \text{in} \quad \mathfrak{F}.$$

Since this relation expresses the orthogonality of $\Phi - \Psi_o$ to all of $\mathfrak{F}$ we realize that Ψ_o is the orthogonal projection

$$\Psi_o = P\Phi$$

of Φ on $\mathfrak{F}$. Thus the projection theorem is proved.

The uniqueness of the orthogonal projection was already established in Section 12.

Another fact, also mentioned in Section 12, should be recalled: Every vector Φ which possesses a projection on a subspace can be written as the sum of a vector $P\Phi$ in it and one, $(1-P)\Phi$, orthogonal to this space. We may amplify this statement now. We denote by $\mathfrak{F}^{\perp}$ the space of all vectors in $\mathfrak{B}$ which are perpendicular to the complete space $\mathfrak{F}$ i.e., the "orthogonal complement" of $\mathfrak{F}$. As an immediate consequence of the projection theorem we then have the

<u>Corollary</u>: Every vector Φ in $\mathfrak{B}$ can be written as the sum

$$\Phi = P\Phi + (1-P)\Phi$$

of a vector in a complete subspace $\mathfrak{F}$ and one in its orthogonal complement $\mathfrak{F}^{\perp}$.

We may express this fact also by saying that the linear combination of the vectors in the complete sub-space $\mathfrak{F}$ and those in $\mathfrak{F}^{\perp}$ span the whole space $\mathfrak{B}$. This fact is symbolically expressed in the form

$$\mathfrak{F} \oplus \mathfrak{F}^{\perp} = \mathfrak{B}$$

We said before that we shall in general deal with cases in which the space $\mathfrak{B}$ itself is complete; we then call it $\mathfrak{H}$. We note: the orthogonal complement $\mathfrak{F}^{\perp}$ of a complete subspace $\mathfrak{F}$ of a Hilbert space is also complete. For, any Cauchy sequence in $\mathfrak{F}$ has a limit in $\mathfrak{H}$, which is also orthogonal to $\mathfrak{F}$, and hence in $\mathfrak{F}^{\perp}$.

Frequently, we shall deal with incomplete subspaces $\mathfrak{F}'$ of a Hilbert space $\mathfrak{H}$. Before the projection theorem can be applied the space $\mathfrak{F}'$ must first be "closed".

A subspace $\mathfrak{F}$ of a normed space $\mathfrak{N}$ is "closed" if every limit of elements in $\mathfrak{F}$ belong to $\mathfrak{F}$. Here we mean by limit a limit element in $\mathfrak{N}$. The closure of an incomplete subspace $\mathfrak{F}'$ of $\mathfrak{N}$ is obtained by joining to $\mathfrak{F}'$ all limit elements.

If the space $\mathfrak{H}$ is complete, every closed subspace of it is complete, as easily verified. The process of closure is then the same as the process of completion. But this process of closure is much simpler than the process of completion described in Section 14 since the elements to be added in a closure process are already available, and the linear combination and the inner product are already defined.

As an application of these considerations we make the following

Remark: Let $\{\Omega^\nu\}$ be an orthonormal system in a Hilbert space $\mathfrak{H}$ and suppose that no vector in $\mathfrak{H}$, except 0, is orthogonal to all unit vectors Ω^ν. Then the system $\{\Omega^\nu\}$ spans the space $\mathfrak{H}$ densely.

In other words, the space $\mathfrak{F}'$ spanned by the vectors Ω^ν, i.e. the space of their finite linear combinations, is dense in $\mathfrak{H}$.

To prove this statement we may consider the closure $\mathfrak{F}$ of $\mathfrak{H}'$, and its orthogonal complement $\mathfrak{F}^\perp$. Then $\mathfrak{H} = \mathfrak{F} \oplus \mathfrak{F}^\perp$

But, by hypothesis, $\mathfrak{F}^\perp$ contains only the zero vector: hence $\mathfrak{H} = \mathfrak{F}$. Thus, it follows that $\mathfrak{F}'$ is dense in $\mathfrak{H}$.

If in the formulation of this remark one drops the requirement that the space $\mathfrak{H}$ be complete, the statement would not necessarily be true. There are counter-examples.

This fact played a considerable role in the earlier theory of integral equations in which one did not require the underlying function space to be complete. A system $\{\Omega^\nu\}$ as described in the Remark was then called "complete"; it was called "closed" if it spanned the function space densely. These two notions were not equivalent. But they are equivalent if the underlying space is complete and then the discrepancy disappears.

It may be felt desirable to have examples presented in which concrete closed subspaces and the projections on them are exhibited; but such examples will not be given here. One may just as well be satisfied with the assertion that the projection theorem will be used over and over again in the course of our presentation of spectral theory.

The subject matter treated in the next section will give an indication of this fact.

18. Bounded Forms

A "linear form" $\chi(\Phi)$ is an assignment of a complex number to every vector Φ of a normed space $\mathfrak{N}$; it is <u>bounded</u> if there is a number χ_o such that

$$|\chi(\Phi)| \leq \chi_o ||\Phi|| \tag{18.1}$$

for every Φ in $\mathfrak{N}$.

A simple example of a linear form of Φ in an inner product space is the inner product (Λ,Φ) of Φ with a fixed vector Λ; this form is bounded by virtue of the Schwarz inequality

$$|\Lambda,\Phi| \leq ||\Lambda|| \; ||\Phi||. \tag{18.2}$$

If the form is defined in a complete inner product space the converse is true:

<u>Theorem 18.1</u>. Let $\chi(\Phi)$ be a bounded linear functional defined in a Hilbert space $\mathfrak{H}$; then there is a vector Λ in $\mathfrak{H}$ such that

$$\chi(\Phi) = (\Lambda,\Phi). \tag{18.3}$$

The proof follows immediately from the projection theorem. Let $\mathfrak{F}$ be the subspace of all those vectors Ψ in $\mathfrak{H}$ for which $\chi(\Psi) = 0$. This space is closed; for, if $\Psi^\sigma \to \Psi_o$ with Ψ^σ in $\mathfrak{F}$ and Ψ_o in $\mathfrak{H}$, we have $|\chi(\Psi_o)| = |\chi(\Psi^\sigma - \Psi_o)|$ $\leq \chi_o||\Psi^\sigma - \Psi_o|| \to 0$, hence $\chi(\Psi_o) = 0$. Therefore, Ψ_o is in $\mathfrak{F}$.

If the space $\mathfrak{F}$ is the full Hilbert space $\mathfrak{H}$, we may set $\Lambda = 0$. Otherwise, by virtue of the Corollary to the Projection Theorem, there is a vector $X_o \neq 0$ in the orthogonal complement

$\mathfrak{F}^{\perp}$ of $\mathfrak{F}$. Clearly, $\chi(X_o) \neq 0$ for such a vector; for otherwise X_o would be in $\mathfrak{F}$ as well as in $\mathfrak{F}^{\perp}$ and would hence be the zero vector.

We now maintain that the space $\mathfrak{F}^{\perp}$ is one-dimensional. In other words, we maintain that every vector X in $\mathfrak{F}^{\perp}$ is a multiple of the vector X_o; specifically, we claim that

$$X = [\chi(X_o)]^{-1}\chi(X)X_o.$$

Clearly, the difference of these two vectors annihilates the form χ; hence, being in $\mathfrak{F}^{\perp}$, this difference is zero. Thus we see that the space $\mathfrak{F}^{\perp}$ is indeed one-dimensional.

By the corollary to the projection theorem every vector in $\mathfrak{H}$ can be written in the form

$$\Phi = \Phi_t + cX_o$$

where Φ_t is in $\mathfrak{F}$ and $c = (X_o,\Phi)/(X_o,X_o)$. Since $\chi(\Phi_t) = 0$ we have

$$\chi(\Phi) = c\chi(X_o);$$

setting

$$\Lambda = (X_o,X_o)^{-1}\chi(X_o)X_o \tag{18.4}$$

we find the desired relation

$$\chi(\Phi) = (\Lambda,\Phi).$$

Theorem 18.1 is thus proved.

As an immediate consequence of Theorem 18.1 we shall prove a corollary concerning bounded (bilinear) forms. A "bilinear form" $\Phi'\underline{B}\Phi$ is an assignment of a complex number to two vectors Φ,Φ' in a space $\mathfrak{B}$ which is linear in Φ and anti-linear in Φ'; it is bounded if there is a number b_o such that

$$(18.5) \qquad |\Phi'\underline{B}\Phi| \leq b_o||\Phi'|| \, ||\Phi||$$

for all Φ,Φ' in $\mathfrak{B}$.

Corollary to Theorem 18.1. Let $\Phi'\underline{B}\Phi$ be a bounded form defined in a Hilbert space $\mathfrak{H}$. Then there is a bounded linear operator B defined in $\mathfrak{H}$ such that the relation

$$(18.6) \qquad \Phi'\underline{B}\Phi = (\Phi',B\Phi)$$

holds for all Φ,Φ' in $\mathfrak{H}$. Furthermore, the validity of the boundedness relation (18.5) for all Φ,Φ' in $\mathfrak{H}$ implies the validity of the relation

$$(18.7) \qquad ||B\Phi|| \leq b_o||\Phi|| \quad \text{for all } \Phi \text{ in } \mathfrak{H}.$$

To prove this corollary we keep Φ fixed and observe that then $\overline{\Phi'\underline{B}\Phi}$ is a bounded linear form in Φ'. Hence, by virtue of Theorem 18.1, there exists a vector Λ in $\mathfrak{H}$ such that $\overline{\Phi'\underline{B}\Phi} = (\Lambda,\Phi')$ or

$$\Phi'\underline{B}\Phi = (\Phi',\Lambda) \quad \text{for all } \Phi' \text{ in } \mathfrak{H}.$$

Obviously, Λ is uniquely determined since the relation

$(\Phi',\Lambda_1) = (\Phi',\Lambda_2)$ for all Φ' in $\mathfrak{H}$ implies $\Lambda_1 = \Lambda_2$.

The relation $\Phi'\underline{B}\Phi^{(1)} = (\Phi',\Lambda^{(1)})$ and $\Phi'\ \underline{B}\Phi^{(2)} = (\Phi',\Lambda^{(2)})$ implies the relation $\Phi'\underline{B}(c_1\Phi^{(1)} + c_2\Phi^{(2)}) = c_1\Phi'\underline{B}\Phi^{(1)} + c_2\Phi'\underline{B}\Phi^{(2)}$ $= (\Phi', (c_1\Lambda^{(1)} + c_2\Lambda^{(2)}))$ for all Φ' in $\mathfrak{H}$ and hence $c_1\Lambda^{(1)} + c_1\Lambda^{(2)} = \Lambda$ by virtue of the uniqueness of Λ just derived. Thus, the vector Λ depends linearly on Φ; the assignment Λ to Φ constitutes a linear operator B.

From the relation (18.6) thus established, combined with (18.5) we derive the relation

$$|\Phi',B\Phi| \leq b_o||\Phi'|| \ ||\Phi||.$$

Setting $\Phi' = B\Phi$ we find $||B\Phi||^2 \leq b_o||B\Phi|| \ ||\Phi||$ and thus (18.7).

Because of the close relationship between bilinear forms and operators we shall not discuss specific bilinear forms now. We shall do so in connection with the discussion of specific operators in the next chapter and again in Chapter VI.

CHAPTER IV

BOUNDED OPERATORS

19. Operator Inequalities, Operator Norm, Operator Convergence

An operator B acting in a Hilbert space $\mathfrak{H}$ was called bounded in Section 6 if there is a number $b > 0$ such that

$$(19.1) \qquad ||B\Phi|| \leq b||\Phi|| \quad \text{for all } \Phi \text{ in } \mathfrak{H}.$$

The bilinear form $(\Psi,B\Phi)$ defined for Φ,Ψ in $\mathfrak{H}$ was called bounded if there is a number $b' > 0$ such that

$$(19.2) \qquad |\Psi,B\Phi| \leq b'||\Psi|| \; ||\Phi|| \quad \text{for all } \Phi,\Psi \text{ in } \mathfrak{H}.$$

Every bound for the bilinear form is one for the operator, and vice versa. For setting $\Psi = B\Phi$ in (19.2) we obtain $||B\Phi||^2 \leq b'||B\Phi|| \; ||\Phi||$ and hence $||B\Phi|| \leq b'||\Phi||$, and by Schwarz's inequality together with (19.1) we obtain $|\Psi,B\Phi| \leq ||\Psi|| \; b||\Phi||$.

The least upper bound for the operator B is denoted by $||B||$; i.e.

$$(19.3) \qquad ||B|| = \text{l.u.b. } ||B\Phi||/\;||\Phi|| \quad \text{for } \Phi \neq 0.$$

This notation anticipates the fact, discussed later on, that the least bound may serve as a "norm".

The statement made above shows that the bound $||B||$ is at the same time the least bound for the form $(\Psi,B\Phi)$; i.e.

$$(19.4) \qquad ||B|| = \text{l.u.b. } |\Psi,B\Phi|/\;||\Phi||\cdot||\Psi|| \quad \text{for } \Phi,\Psi \neq 0.$$

An operator B^* acting in the space $\mathfrak{H}$ was called the "formal adjoint" of the bounded operator B if the relation

(19.5) $\quad (B\Phi,\Psi) = (\Phi,B^*\Psi)$ holds for all Φ,Ψ in $\mathfrak{H}$.

Later on, in Chapter VI, we shall introduce a "strict" adjointness property and show that every formal adjoint of a bounded operator is the strict adjoint. For this reason we shall drop the qualification "formal" in this chapter.

It is an important fact that <u>to every bounded operator B acting in a Hilbert space there is just one such adjoint operator</u>. Since $(B\Phi,\Psi)$ is a bounded bilinear form the existence of B^* follows immediately from the corollary to Theorem 18.1; its uniqueness is obvious since for the difference $\tilde{B}$ of two adjoint operators the relation $(\Phi,\tilde{B}\Psi) = 0$ holds for all Φ,Ψ, whence $\tilde{B}\Psi = 0$ for all Ψ, i.e. $\tilde{B} = 0$.

Since the least bound $||B||$ of the operator B is at the same time the least bound of the form $(\Psi,B\Phi)$ and $(\Phi,B^*\Psi) = \overline{(\Psi,B\Phi)}$, it is clear that $||B||$ is also the bound of the operator B^*,

(19.6) $$||B^*|| = ||B||.$$

The bounded operator B will be called "self-adjoint" (without the qualification "formal", see Section 9) if it is equal to its adjoint, $B = B^*$.

The statement that the least bound of an operator is the same as the least bound of the associated bilinear form can be strengthened for self-adjoint operators. To such an operator B we assign its "quadratic form" $(\Phi,B\Phi)$.

Note that the bilinear form $(\Psi,B\Phi)$ of an operator B is determined by the quadratic form $(\Phi,B\Phi)$ if B is self-adjoint: this is seen from the identity

$$2(\Psi,B\Phi) + 2(\Phi,B\Psi) = ((\Psi+\Phi,\ B(\Psi+\Phi)) - ((\Psi-\Phi,\ B(\Psi-\Phi)).$$

The left member equals $4(\Psi,B\Phi)$ if the space is real. If the space is complex we need only add to this identity the identity obtained from it by substituting $i\Psi$ for Ψ.

The quadratic form $(\Phi,B\Phi)$ of the self-adjoint operator B will be called "non-negative" if the inequality

$$(19.7) \qquad (\Phi,B\Phi) \geq 0 \quad \text{holds for} \quad \Phi \quad \text{in} \quad \mathfrak{H}.$$

(Note that the value of this form is real for Hermitean B.) We use the notation

$$B \geq 0$$

to express this property.

The bilinear form $(\Psi,B\Phi)$ of such a non-negative self-adjoint operator may be considered a semi-inner-product since it satisfies the requirements (7.1), (7.2), and (7.3). Therefore, the Schwarz inequality

$$(19.8) \qquad |\Psi,B\,\Psi|^2 \leq (\Psi,B\Phi)\ (\Phi,B\Phi) \qquad \text{if} \quad B \geq 0$$

holds with any Φ,Ψ in $\mathfrak{H}$.

Using it we may derive the general

<u>Theorem 19.1</u>. Suppose the quadratic form $(\Phi,B\Phi)$ is non-negative and has the bound b, i.e. $0 \leq B \leq b$, or

(19.9) $$0 \leq (\Phi, B\Phi) \leq b||\Phi||^2.$$

Then the inequality

(19.10) $$||B\Phi||^2 \leq b(\Phi, B\Phi) \quad \text{holds.}$$

In fact, we need only set $\Psi = B\Phi$ in (19.8), and apply (19.9) to $B\Phi$ instead of Φ, and divide by $||B\Phi||^2$. If $||B\Phi|| = 0$ relation (19.10) holds anyway.

Next we state the important

Theorem 19.2. Suppose the quadratic form of the bounded Hermitean operator B has the bound b,

(19.11) $$|\Phi, B\Phi| \leq b||\Phi||^2;$$

then b is a bound for the operator B

(19.12) $$||B\Phi|| \leq b||\Phi||.$$

In case the inequality

$$|\Phi, B\Phi| < b||\Phi||^2$$

holds for all $\Phi \neq 0$, also the inequality

$$||B\Phi|| < b||\Phi|| \quad \text{for} \quad \Phi \neq 0$$

holds.

Condition (19.11) may be expressed by saying that the forms $b \pm B$ are non-negative, $b \pm B \geq 0$, or,

$$(19.11)' \qquad -b \leq B \leq b$$

Theorem 19.2 can then be stated as saying that (19.11)' implies

$$(19.1)' \qquad ||B|| \leq b.$$

There are many ways of proving Theorem 19.2. One concise proof is based on the identity

$$2b(b^2-B^2) = (b-B)(b+B)(b-B) + (b+B)(b-B)(b+B)$$

and the resulting identity

$$2b[b^2(\Phi,\Phi) - (B\Phi,B\Phi)]$$
$$= ((b-B)\Phi, (b+B)(b-B)\Phi) + ((b+B)\Phi, (b-B)(b+B)\Phi).$$

The right hand side is non-negative since the operators $b \pm B$ are non-negative. Hence the left hand side is non-negative. This implies (19.1)' unless $b = 0$. If $b = 0$, inequality (19.11) holds for any positive b_1; hence so does (19.12). Since $b_1 > 0$ is arbitrary in this case, (19.1)' holds for $b = 0$. The statement involving $<$ in place of $\leq$ is proved in the same way.

Incidentally, the last result: $B = 0$ if $(\Phi,B\Phi) = 0$ for all Φ, holds even if the operator is not Hermitean, provided the space $\mathfrak{H}$ is complex.

The least bound $||B||$ of a bounded operator may serve as a norm in the linear space of all bounded operators. For, postulates (6.1) to (6.3) are evidently satisfied and (6.4), i.e. the triangle inequality

$$||B_1 + B_2|| \leq ||B_1|| + ||B_2|| \tag{19.13}$$

follows immediately from the relation $||(B_1 + B_2)\Phi||$ $\leq \{||B_1|| + ||B_2||\}||\Phi||$, which states that $||B_1|| + ||B_2||$ is an upper bound for $B_1 + B_2$. For this reason the least bound $||B||$ is also called the "minimal" norm of B, sometimes simply called the "norm". For restricted classes of operators we shall on occasion, see Section 20, use other - non-minimal - norms, given by other than least bounds.

Naturally, a notion of convergence of operators can be introduced with the aid of the norm $||\ ||$: We say, a sequence of bounded operators B^ν tends to a bounded limit operator "in minimal norm" or "uniformly" if

$$||B^\nu - B|| \to 0 \quad \text{as} \quad \nu \to \infty. \tag{19.14}$$

Convergence in any other (non-minimal) norm may also be introduced; convergence with respect to a non-minimal norm is stronger than uniform convergence.

We shall frequently use the notion of convergence in norm; but for many purposes a weaker kind of convergence, called "strong" convergence will be more suitable since a sequence may converge strongly even if it does not converge in norm and important conclusions can often be deduced from strong convergence.

We say, a sequence of bounded operators B^ν converges strongly to a bounded operator B if

(1) the vectors $B^\nu\Phi$ converge to $B\Phi$ for every vector Φ in $\mathfrak{H}$,

$$||B^\nu\Phi - B\Phi|| \to 0 \quad \text{as } \nu \to \infty$$

(2)* the operators B are uniformly bounded; i.e. there is a number b_o such that

$$||B^\nu|| \leq b_o \quad \text{for all } \nu.$$

Clearly, the norm $||B||$ of the limit operator B is bounded by any common bound of the sequence $||B^\nu||$ or a sub-sequence thereof. This sub-sequence may be so chosen as to approach the inferior limit $\inf_\nu ||B^\nu||$ of the sequence $||B^\nu||$. From this one derives the inequality

$$(19.15) \qquad ||B|| \leq \inf_\nu ||B^\nu||$$

for the strong limit B of the sequence B^ν.

We also mention the important, though immediately verified fact that <u>the products</u> $B_1^\nu B_2^\nu$ <u>of the members of two strongly convergent sequences form again a strongly convergent sequence</u>.

A sequence of bounded operators B^ν is a <u>strong</u> Cauchy sequence if

(1) the vectors $B^\nu\Phi$ form a Cauchy sequence for each vector in $\mathfrak{H}$.

(2) the operators B_ν are uniformly bounded, $||B^\nu|| \leq b_o$.

---

*We mention incidentally that condition (2) could be omitted since the existence of a uniform bound b_o could be deduced from condition (1) by virtue of an extension of the theorem mentioned in the footnote on page 145. We shall not have occasion to use this remarkable fact since the procedure by which we shall introduce our sequences of bounded operators will always automatically yield a uniform bound for them.

Such a sequence evidently possesses a limit operator B to which it strongly converges and for which b_o is a bound; $||B|| \leq b_o$.

Clearly, the product of two strongly convergent Cauchy sequences is again a strongly convergent Cauchy sequence.

Finally, we mention the notion of "weak" convergence of a sequence of uniformly bounded operators B^ν; by this it is simply meant that the bilinear forms $(\Psi, B^\nu \Phi)$ converge for all Φ, Ψ. The main weakness of this type of convergence is that the product of two weakly convergent operators need not converge weakly.

There is one particular case of weak convergence of operators which automatically implies strong convergence, as seen from

<u>Theorem 19.3 on monotone convergence</u>. Let $B_1, B_2, \ldots$ be a sequence of bounded Hermitean operators which increases monotonically,

$$B_1 \leq B_2 \leq B_3 \leq \cdots$$

and is uniformly bounded, $||B_\tau|| \leq b$, $\tau = 1,2,\ldots$. Then the sequence B_τ is a strong Cauchy sequence.

The assumption $B_\sigma \leq B_\tau$ for $\sigma \leq \tau$ implies that $(\Phi, (B_\tau - B_\sigma)\Phi) \geq 0$ for all Φ. Since evidently $||B_\tau - B_\sigma|| \leq 2b$ we may apply Theorem 19.1 which gives

$$||(B_\tau - B_\sigma)\Phi||^2 \leq 2b(\Phi, (B_\tau - B_\sigma)\Phi).$$

Since the sequence $(\Phi, B_\tau\Phi)$ increases monotonically and is bounded, the sequence $(\Phi, (B_\tau - B_\sigma)\Phi)$ tends to zero as $\sigma, \tau \to \infty$. Hence the statement follows.

Inasmuch as a strong Cauchy sequence of bounded operators leads to a limit operator, this limit process may be used to define specific operators with the aid of operators already defined before. We shall use this procedure extensively. Before doing so, however, we shall describe another procedure of defining an operator, the extension of a bounded operator defined in a dense subspace of $\mathfrak{H}$. Accordingly, we formulate the rather obvious

Second Extension Theorem. Suppose the operator B is defined in a dense subspace $\mathfrak{H}'$ of $\mathfrak{H}$, produces vectors in $\mathfrak{H}$, and is bounded

$$||B\Phi|| \leq b||\Phi|| \quad \text{for} \quad \Phi \quad \text{in} \quad \mathfrak{H}'.$$

Then there exists an operator defined in all of $\mathfrak{H}$, having the bound b there, and agreeing in $\mathfrak{H}'$ with B.

Every Φ in $\mathfrak{H}$ can be approximated by a sequence Φ^{ν} from $\mathfrak{H}'$. By virtue of $||B\Phi^{\nu} - B\Phi^{\mu}|| \leq b||\Phi^{\nu} - \Phi^{\mu}||$ the vectors $B\Phi^{\nu}$ form a Cauchy sequence. The limit is evidently independent of the choice of the vectors Φ^{ν}; we may denote it by $B\Phi$ since it evidently agrees with $B\Phi$ if Φ is in $\mathfrak{H}'$. The linearity of the operator B so defined is obvious and also the relation $||B\Phi|| \leq b||\Phi||$.

20. Integral Operators

Specific cases of bounded operators are naturally given by integral operators. These are operators which act on the functions of some function space and produce functions in this space.

For example, let the functions be the continuous functions $\phi(s)$ defined in a closed interval $\mathscr{I}$ of the s-axis. Then an integral operator K may be given in the form

$$K\phi(s) = \int_{\mathscr{I}} k(s,s')\phi(s')ds' \tag{20.1}$$

with the aid of a function $k(s,s')$ of two variables, both running over the interval $\mathscr{I}$. This function $k(s,s')$ is called the "kernel" of the integral operator. For the present let us assume that this kernel is continuous over the square $\mathscr{I} \times \mathscr{I}$. Then the function $K\phi(s)$ is also continuous. Moreover, we have

$$\max_s |K\phi(s)| \leq \underline{k} \max_s |\phi(s)|$$

with

$$\underline{k} = I \max_{s,s'} |k(s,s')| , \tag{20.2}$$

where I is the length of the interval $\mathscr{I}$. Consequently we have

$$||K\phi|| \leq \underline{k} \; ||\phi|| . \tag{20.3}$$

no matter whether we take the maximum norm or the square integral norm. In any case the operator K is bounded.

Still we cannot immediately apply to the operator K the general theory of bounded operators which we have begun to develop. For in this theory it is assumed that the bounded operators are defined in the whole complete Hilbert space; but the space of continuous functions is not complete with respect to the square integral norm.

The obstacle we thus have met can be easily overcome. We need only employ the second extension theorem described at the end of Section 19. This is possible since the space of continuous functions is dense in its completion. According to this theorem there does exist a bounded operator acting on all (ideal) functions $\phi(s)$ in the Hilbert space with the unit form

$$(\phi,\phi) = \int_{\mathscr{S}} |\phi(s)|^2 ds, \tag{20.4}$$

which agrees with the given integral operator. We denote this extended operator by K; in fact we shall use the formula (20.1) to describe it symbolically.

It is necessary for us to introduce a more general class of integral operators. On the one hand we must consider a more general function space in which the operator acts, and on the other hand we must consider a more general class of kernels. In defining integral operators we shall employ both, the second extension theorem, and approximation by a Cauchy sequence of operators already defined. In this connection we shall have to use bounds for the operators which are less crude than the bound $\underline{k}$ given by (20.2).

It seems advisable to discuss such less crude bounds already for the simple integral operator (20.1) with a continuous kernel acting on functions with the unit form (20.4), although for this simple integral operator as such we do not need these bounds.

Using a number α between zero and 1

$$0 \leq \alpha \leq 1, \tag{20.5}$$

we use Schwarz's inequality to estimate

$$|K\phi(s)|^2 \leq \left| \int_{\mathscr{S}} |k(s,s')|^{\alpha} |k(s,s')|^{1-\alpha} |\phi(s')| ds' \right|^2$$

$$\leq \int_{\mathscr{S}} |k(s,s')|^{2\alpha} ds' \int_{\mathscr{S}} |k(s,s'')|^{2-2\alpha} |\phi(s'')|^2 ds''$$

whence, after interchanging the order of integration,

$$\int_{\mathscr{S}} |K\phi(s)|^2 ds \leq \int_{\mathscr{S}} q_{2\alpha}^2(s'')|\phi(s'')|^2 ds''$$

with

$$(20.6) \qquad q_{2\alpha}(s'') = \left| \int_{\mathscr{S}} \left| \int_{\mathscr{S}} |k(s,s')|^{2\alpha} |k(s,s'')|^{2-2\alpha} ds \right| ds' \right|^{1/2} .$$

Using the abbreviation

$$(20.7) \qquad k_{2\alpha} = \underset{s''}{\text{l.u.b.}}\ q_{2\alpha}(s'')$$

we evidently obtain the estimate

$$(20.8) \qquad \int_{\mathscr{S}} |K\phi(s)|^2 ds \leq k_{2\alpha}^2 \int_{\mathscr{S}} |\phi(s)|^2 ds,$$

or

$$(20.8)' \qquad ||K\phi|| \leq k_{2\alpha}\ ||\phi||.$$

For $\alpha = 1$ the bound $k_{2\alpha}$ becomes the "Hilbert-Schmidt" bound

$$(20.7)_2 \qquad k_2 = \left| \int_{\mathscr{S}} \int_{\mathscr{S}} |k(s,s')|^2 ds ds' \right|^{1/2} ;$$

for $\alpha = 1/2$ we obtain the bound

$$(20.7)_1 \qquad k_1 = \left| \underset{s''}{\text{l.u.b.}} \int_{\mathscr{S}} \int_{\mathscr{S}} |k(s,s')|\ |k(s,s'')| ds ds' \right|^{1/2} ,$$

related to the "Holmgren bound" to be discussed below.

We now consider a general measure function $r(s)$ and the space $\mathfrak{H}$ of functions $\phi(s)$ complete with respect to the unit form

(20.10) $$(\phi,\phi) = \int |\phi(s)|^2 dr(s).$$

We recall that this space was the completion of piecewise constant functions with bounded support. We enlarged this subspace first by adjoining the general piecewise continuous functions with bounded support, then by adjoining those piecewise continuous functions defined for $-\infty < s < \infty$ for which the unit form is finite, and finally, by adjoining the ideal functions needed to complete the space.

In order to define integral operators we first introduce piecewise constant kernels. We introduce the product $P \times P$ of a partition of the s-axis and the same partition of the s'-axis. The cells of this partition are open rectangles, open segments in the s- or s'-direction, and points. A function $k(s,s')$ will be called piecewise constant if it is constant on the cells of such a product partition; the function will be called piecewise continuous if on each product cell it agrees with a function which is continuous on the closure of this cell. With the aid of such kernels $k(s,s')$ we define the integral operator K through

(20.11) $$K\phi(s) = \int k(s,s')\phi(s')dr(s').$$

First we assume ϕ to be of bounded support; $\phi \in \dot{\mathfrak{C}}$. If ϕ and k are piecewise constant, then so is $K\phi(s)$; if ϕ and k are piecewise continuous, then so is $K\phi(s)$. So far we leave open whether or not $\int |K\phi(s)|^2 dr(s) < \infty$.

We now introduce bounds for the kernel $k(s,s')$. We fix a number α from 0 to 1, introduce the number

$$(20.12) \quad k_{2\alpha} = \left| \underset{s''}{\text{l.u.b}} \int\int |k(s,s')|^{2\alpha} |k(s,s'')|^{2-2\alpha} dr(s)\,dr(s') \right|^{1/2} .$$

and require of the kernel $k(s,s')$ that for it

$$(20.13) \qquad k_{2\alpha} < \infty.$$

A literal repetition of the arguments that led to formula (20.8) now leads to the inequality

$$(20.14) \qquad \int |K\phi(s)|^2 dr(s) \leq k_{2\alpha}^2 \int |\phi(s)|^2 dr(s)$$

for piecewise continuous functions of bounded support. This result shows that the function $K\phi(s)$ is in the Hilbert space $\mathfrak{H}$. We may write (20.14) in the form

$$(20.14)' \qquad ||K\phi|| \leq k_{2\alpha} ||\phi||.$$

Since the space $\dot{\mathfrak{C}}$ of piecewise continuous functions of bounded support is dense in the Hilbert space $\mathfrak{H}$ we may apply our second extension theorem. The operator K is then defined in all of $\mathfrak{H}$, and inequality (20.14)' holds there. We shall use relation (20.11) as the symbolic description of this operator.

For later purposes it is desirable to introduce instead of $k_{2\alpha}$ a different bound which (except for $\alpha = 2$) is weaker than the bound $k_{2\alpha}$ but has the advantage that it may serve as a norm in a space of kernels.

First we observe that k_2 is evidently a square integral norm in the space of these piecewise continuous functions for which it is finite. We call this space $\mathfrak{H}_2'$ and adopt

$$(20.15) \qquad ||K||_2 = k_2$$

as norm in it. We call $||K||_2$ the Hilbert-Schmidt norm.

For $2\alpha = 1$ we introduce the quantity

$$(20.16) \quad ||K||_1 = \max\left\{ \underset{s^*}{\text{l.u.b.}} \int |k(s^*,s')|dr(s'),\ \underset{s''}{\text{l.u.b.}} \int |k(s,s'')|dr(s) \right\}.$$

As easily verified, we have

$$(20.17) \qquad ||K||_1 \geq k_1;$$

hence the operator K is defined whenever $||K||_1 < \infty$. Also $||K||_1$ satisfies the triangle inequality since evidently

$$\int |k_1(s^*,s') + k_2(s^*,s')|dr(s'),\ \int |k_1(s,s'') + k_2(s,s'')|dr(s)$$
$$\leq ||K_1||_1 + ||K_2||_1 ;$$

at the same time it follows that the set $\mathfrak{H}_1'$ of piecewise continuous kernels $k(s,s')$ with $||K||_1 < \infty$ forms a linear space. Therefore, $||K||_1$ may serve as a norm. We call $||K||_1$ the "Holmgren norm".

Of course, other norms could be introduced instead of the ones described. For example, employing

$$q_1^{(\alpha)}(s'') = \left| \int_{\mathscr{S}} \left| \int_{\mathscr{S}} |k(s,s')|a(s')|k(s,s'')|a^{-1}(s'')dr(s) \right| dr(s') \right|^{1/2}$$

in place of q_1, one easily verifies the observation made by Carleman that instead of $||K||_1$ the expression

$$(20.18)\quad ||K||_{1,a,b} = \max\left\{ \underset{s^*}{\text{l.u.b.}}\ b^{-1}(s^*) \int |k(s^*,s')|a(s')dr(s'),\ \underset{s''}{\text{l.u.b.}}\ a^{-1}(s'') \int |k(s,s'')|b(s)dr(s) \right\},$$

involving two positive functions $a(s)$, $b(s)$ may be taken as norm.

An application may be mentioned incidentally. Setting $a(s'') = \int |k(s,s'')|dr(s)/k_1^*$ and $b(s) = 1$ one obtains the k_1-bound k_1^* of the adjoint operator K^* as bound for K.

Obviously there are kernels for which $||K||_2$ is finite and $||K||_1$ infinite or the other way around. Examples could be furnished by product kernels

$$k(s,s') = g(s)h(s').$$

Evidently

$$||K||_2 = \left| \int |g(s)|^2 dr(s) \int |h(s')|^2 dr(s') \right|^{1/2}$$

and

$$||K||_1 = \max\left\{ \underset{s^*}{\text{l.u.b.}}\ |g(s^*)| \int |h(s')|dr(s'),\ \underset{s''}{\text{l.u.b.}}\ |h(s'')| \int |g(s)|dr(s) \right\}.$$

Any one of the quantities entering here could be made infinite while the others are finite.

Important kernels for which the Holmgren norm is appropriate are kernels of the form

$$k(s,s') = g(s-s').$$

Here we assume $r(s) = s$, $-\infty < s < \infty$. Evidently, always $||K||_2 = \infty$ unless $g(t) \equiv 0$; but

$$||K||_1 = \int_{-\infty}^{\infty} |g(t)|dt \tag{20.19}$$

may well be finite.

An example for which the Holmgren norm is not sufficient but Carleman's modification is sufficient is given by the integral operator (related to the "Hilbert matrix")

$$K\phi(s) = \int_1^{\infty} \phi(s') \frac{ds'}{s+s'}$$

which acts in the space of functions $\phi(s)$ for $s \geq 1$ with $(\phi,\phi) = \int_1^{\infty} |\phi(s)|^2 ds$. Here the Carleman norm with $a(s) = b(s) = 1/\sqrt{s}$ is finite; in fact $||K||_{1,a,a} = \pi/2$, as a simple computation shows.

Finally we mention that the class of integral operators we have introduced includes <u>infinite matrices</u> acting on sequences of numbers $\xi = \{\xi_\sigma\}$ for which the infinite series

$$(\xi,\xi) = \sum_\sigma |\xi_\sigma|^2$$

is finite, i.e. on vectors in the special Hilbert space. The operator K, given by

$$K\xi = \left\{ \sum_{\sigma'} k_{\sigma\sigma'} \, , \, \xi_{\sigma'} \right\},$$

is defined and bounded if any of the norms $k_{2\alpha}$ is finite, such as

$$||K||_2 = \left| \sum_{\sigma,\sigma'} |k_{\sigma\sigma'}|^2 \right|^{1/2}$$

or

$$||K||_1 = \max\left\{ \underset{\sigma^*}{\text{l.u.b.}} \sum_{\sigma'} |k_{\sigma^*\sigma'}|, \quad \underset{\sigma''}{\text{l.u.b.}} \sum_{\sigma} |k_{\sigma\sigma''}| \right\}.$$

The kernels $k(s,s')$ so far considered were assumed to be piecewise continuous in the specific sense explained above. It is necessary to extend the class of these kernels and the class of operators generated by them. To this end one may adopt a norm in an appropriate space of kernels and then complete this space with respect to this norm. Clearly, we can adopt for this purpose any of the norms $||K||_{2\alpha}$ defined in the space $\mathfrak{H}_{2\alpha}$ of piecewise continuous kernels. We then can close off these spaces by adjoining "ideal kernels" thus obtaining a complete space of such kernels. When a sequence of kernels k^ν in $\mathscr{R}'_{2\alpha}$ approaches a kernel k in $\mathscr{R}_{2\alpha}$ in the sense of the norm, the corresponding operator K^ν approaches a limit operator K, i.e. $||K^\nu - K||_{2\alpha} \to 0$. The ideal function k may be regarded as the kernel of this operator, which therefore may symbolically be written as an integral operator.

This method of defining integral operators may frequently be used in cases in which a kernel is given which is piecewise continuous except at the diagonal $s' = s$, where it may be singular. The "Volterra operator", whose kernel is continuous up to the diagonal on one side, and vanishes on the other side, is such a case.

We exclude the case in which $k(s,s')$ differs from zero at points on the diagonal $s' = s$ where $r(s)$ has a jump. We then introduce an approximate kernel k_ε which vanishes in a neighborhood of the diagonal and agrees with k outside of this neighborhood. Specifically, we assume this neighborhood η_ε to be an open set composed of cells of a partition $\mathscr{P} \times \mathscr{P}$, contained in the strip $|s-s'| \leq 2\varepsilon$ and containing the strip $|s-s'| \leq \varepsilon$.

We first adopt the norm $||K||_2$ and accordingly require that the integral

$$||K||^2 = \int\int_{\mathscr{M}_\varepsilon} |k(s,s')|^2 dr(s)dr(s')$$

over the complement $\mathscr{M}_\varepsilon$ of the neighborhood η_ε remains bounded as $\varepsilon \to 0$. The limit of this integral will then be denoted by

$$\int\int |k(s,s')|^2 dr(s)dr(s').$$

The approximate kernels k_ε evidently form a Cauchy sequence

$$||K_\delta - K_\varepsilon||_2^2 = \iint_{\eta_\delta - \eta_\varepsilon} |k(s,s')|^2 dr(s)dr(s') \to 0 \quad \text{as} \quad \delta > 2\varepsilon \to 0.$$

Therefore, the operators K_ε tend to a limit operator K, represented by the kernel $k(s,s')$. Also we may set

$$||K||_2 = \left| \int\int |k(s,s')|^2 dr(s)dr(s') \right|^{1/2}.$$

Similar arguments may also be used to define operators that have singularities at places other than the diagonal. Also these arguments can be used to show that the operator K_a with the kernel

$$\begin{aligned} k_a(s,s') &= k(s,s') \quad \text{for} \quad |s| \leq a, \quad |s'| \leq a \\ &= 0 \quad \text{for other} \quad s,s', \end{aligned}$$

tends to the operator K, in the Hilbert-Schmidt norm,

$$||K_a - K||_2 \to 0 \quad \text{as} \quad a \to \infty.$$

For the Holmgren norm a corresponding procedure to extend integral operators might not succeed since the norm $||K_\varepsilon||_1$ (or $||K_a||_1$) does not necessarily increase monotonically as $\varepsilon \to 0$ (or $a \to \infty$). Still, the Holmgren norm can be used to extend the class of integral operators, if one is satisfied with using strong convergence of operators, rather than convergence in norm.

What may happen in such a case will be illustrated in a special case which is of considerable importance.

We consider the Hilbert space $\mathfrak{H}$ of functions $\phi(s)$ defined for $-\infty < s < \infty$ and carrying the unit form

$$(\phi,\phi) = \int |\phi(s)|^2 ds.$$

We then consider the integral operators J^τ with the kernel

$$j^\tau(s,s') = \tau j(\tau(s-s')) \tag{20.20}$$

where τ is any positive number and $j(x)$ is a real continuous function of the real variable x with the following properties

(20.21)

1) $j(x) \geq 0$,

2) $j(x) = 0$ for $|x| \geq 1$,

3) $\displaystyle\int_{-1}^{1} j(x)dx = 1.$

The Holmgren norm of these operators is obviously

(20.22) $$||J^\tau||_1 = 1.$$

Clearly, then $j^\tau(s,s') = 0$ for $|s-s'| \geq 1/\tau$ and

$$\int j^\tau(s,s')ds' = 1.$$

Now we maintain that the operators J^τ tend strongly to the identity as $\tau \to \infty$, i.e.,

(20.23) $$||J^\tau\phi - \phi|| \to 0 \quad \text{as} \quad \tau \to \infty \quad \text{for every } \phi \text{ in } \mathfrak{H}.$$

For continuous functions ϕ of bounded support the statement follows from the estimate

$$|J^\tau\phi(s) - \phi(s)| = \left| \int j^\tau(s,s')[\phi(s') - \phi(s)]ds' \right|$$
$$\leq \max_{\tau|s'-s|\leq 1} |\phi(s') - \phi(s)|.$$

We need only note that the last term here can be made arbitrarily small for sufficiently large τ independently of s, by virtue of the uniformity of the continuity of $\phi(s)$.

To an arbitrary ϕ in $\mathfrak{H}$ and any $\varepsilon > 0$ we may find a continuous ϕ_ε with bounded support such that $||\phi_\varepsilon - \phi|| \leq \varepsilon$. Then also $||J^\tau\phi_\varepsilon - J^\tau\phi|| \leq ||J^\tau||\ ||\phi_\varepsilon - \phi|| = ||\phi_\varepsilon - \phi|| \leq \varepsilon$. Hence $||J^\tau\phi - \phi|| \leq ||J^\tau\phi - J^\tau\phi_\varepsilon|| + ||J^\tau\phi_\varepsilon - \phi_\varepsilon|| + ||\phi_\varepsilon - \phi|| \leq 3\varepsilon$. Thus the statement is proved.

Since the identity is thus recognized as the strong limit of a sequence of integral operators it may be suggested to write it symbolically as an integral operator. In fact, this was done by Dirac who introduced the notation $\delta(s-s')$ for this kernel. (The symbolic kernel is regarded as a function of $s - s'$ since the approximating kernels have this property.)

We should note that this delta kernel is <u>not</u> obtained as an ideal function of s and s' by a closure process with respect to a norm in the space of functions $k(s,s')$.

It was said above that the extension of an integral operator in norm corresponds to a "strong" extension of the corresponding kernels to ideal elements in a normed space of kernel functions. We now realize that the extension of kernel functions corresponding to strong extension of an integral operator does not correspond to a strong extension of the kernel; it is of a different character. In fact, it could be described with the aid of the notion of distribution. Later on we shall come back to this question.

21. Functions of Bounded Operators

We now return to the general theory of bounded operators. We shall establish a functional calculus of such operators. To every bounded piecewise continuous function $f(\beta)$ we shall assign an operator $f(B)$ which obeys the rules of functional calculus, described in Chapter I. We assume the self-adjoint operator B to act in a Hilbert space $\mathfrak{H}$ and we let b be a bound for it, which may be taken as the norm $||B||$. In any case

$$||B|| \leq b, \quad \text{or} \quad ||B\Phi|| < b||\Phi|| \quad \text{for } \Phi \text{ in } \mathfrak{H}.$$

Since B acts in all of $\mathfrak{H}$ the operators $B^2, B^3, \ldots$ are defined and so is every polynomial $p(B) = \sum_{\rho=0}^{n} a_\rho B^\rho$ of degree n. At present we assume the coefficients a_ρ to be real. Then we immediately verify that the operator $p(B)$ is self-adjoint just as B is. Our aim is to extend the definition of the operator $f(B)$ from polynomials to functions of a more general class. The main tool in doing this is the

<u>Main Lemma</u>. Let B be a self-adjoint operator in $\mathfrak{H}$ with the bound $||B|| \leq b < \infty$. Let $p(\beta)$ be a real polynomial of the real variable β for which

$$p(\beta) \geq 0 \quad \text{for} \quad |\beta| \leq b.$$

Then

$$p(B) \geq 0.$$

The statement involves the terminology introduced in Section 19; accordingly it means

$$(\Phi, p(B)\Phi) \geq 0 \quad \text{for} \quad \Phi \quad \text{in} \quad \mathfrak{H}.$$

Many proofs of this Main Lemma have been given. We present a proof which goes back to F. Riesz, but uses a modification of Riesz's argument suggested by K. Brokate.

In addition to the class (p) of real polynomials p with $p(\beta) \geq 0$ for $|\beta| \leq b$ we introduce the class (q) of polynomials which are sums of polynomials of the form $q(\beta) = g^2(\beta)$, $(b\mp\beta)g^2(\beta)$ where $g(\beta)$ is any real polynomial. We note that the sum and product of polynomials in (q) belong to (q) again. For, the product of two squares is a square and the product of $(b+\beta)$ with $(b-\beta)$ can be written in the form $(b+\beta)(b-\beta) = \frac{1}{2b}(b+\beta)(b-\beta)^2 + \frac{1}{2b}(b-\beta)(b+\beta)^2$. Evidently, (q) is contained in (p), but the converse is also true:

Lemma on Polynomials $(q) = (p)$.

We need only prove that every polynomial of class (p) belongs to (q). Let $(p)^n$ and $(q)^n$ be the classes of polynomials of degree $\leq n$ in (p) and (q). Then we prove the lemma by induction, assuming that the statement $(q)^{n-1} = (p)^{n-1}$ is true.

Let $p(\beta)$ be a polynomial in $(p)^n$ which vanishes somewhere in $|\beta| < b$. If the polynomial $p(\beta)$ vanishes at a point β_o in the interior, $|\beta_o| < b$, we may write it in the form $p(\beta) = (\beta-\beta_o)^2 p_2(\beta)$ where $p_2(\beta)$ is a polynomial of degree $\leq n-2$. Evidently p_2 belongs to the class $(p)^{n-2}$ since p belongs to $(p)^n$. By induction assumption, therefore, p_2 belongs to $(q)^{n-2}$, hence p belongs to $(q)^n$. If $p(\beta)$ vanishes at $\beta_o = \pm b$ we may write $p(\beta) = (b\mp\beta)p_1(\beta)$. Evidently, p_1 belongs to $(p)^{n-1}$, hence to $(q)^{n-1}$. Therefore p belongs to $(q)^n$. If p in $(p)^n$ does not vanish in $|\beta| \leq b$, it has a positive minimum, p_{min} there, and hence can be written in the form $p(\beta) = p_{min} + p_o(\beta)$ where now $p_o(\beta)$ is in $(p)^n$ and vanishes somewhere in $|\beta| \leq b$. Again we conclude that $p(\beta)$ is in $(q)^n$. Thus $(q)^n = (p)^n$ follows and the lemma on polynomials is proved.

Setting $g(B)\Phi = \Psi$, we can write each form $(\Phi, q(B)\Phi)$ as the sum of terms of the form

$$(\Psi,\Psi),\ (\Psi,(b-B)\Psi),\ (\Psi,(b^2-B^2)\Psi) = b^2(\Psi,\Psi) - (B\Psi,B\Psi).$$

Each of these terms is non-negative, hence we have

$$(\Phi,q(B)\Phi) \geq 0.$$

since every polynomial in (p) is of the form q, the inequality $(\Phi,p(B)\Phi) \geq 0$ follows and the Main Lemma is proved.

As a consequence of the Main Lemma we may state the

<u>Corollary to the Main Lemma</u>. Let the real polynomial $p(\beta)$ be such that

$$|p(\beta)| \leq p_o \quad \text{for} \quad |\beta| \leq b.$$

Then

$$||p(B)|| \leq p_o.$$

Since the polynomials $p_o \mp p(\beta)$ are non-negative in $|\beta| \leq b$, the Main Lemma is applicable to them. It yields the relation $|(\Phi, p(B)\Phi)| \leq p_o$ or $-p_o \leq p(B) \leq p_o$. The statement of the corollary then follows from Theorem 19.2.

We now can prove the

<u>Main Theorem</u>. Let B be a Hermitean operator with the bound b. To every continuous function $f(\beta)$ defined for $|\beta| \leq b$ there can be assigned an operator, denoted by $f(B)$, which obeys the rules I, II, III, IV of operational calculus:

I. $f_1(\beta) + f_2(\beta) = f(\beta)$ implies $f_1(B) + f_2(B) = f(B)$,

II. $f_1(\beta)f_2(\beta) = f(\beta)$ implies $f_1(B)f_2(B) = f(B)$.

III. Whenever the values of the real-valued function $f(\beta)$ lie in an interval $\mathscr{K}$, also the values of the ratio

$$(\Phi, f(B)\Phi)/(\Phi,\Phi) \qquad \text{for any vector} \quad \Phi \neq 0 \quad \text{in} \quad \mathfrak{H}$$

lie in $\mathscr{K}$.

This ratio may be interpreted as a mean value of the operator

$f(B)$ generated by the vector Φ; thus this mean value lies in the same interval to which the function $f(\beta)$ is restricted.

IV. Whenever $f(\beta) = 0$ for $|\beta| \leq b$, the operator $f(B) = 0$.

An immediate corollary to rule III, via Theorem 19.2, is the rule

III'. Let f_o be the center of the interval $\mathcal{K}$ and 2δ its diameter; then

1) $||(f(B) - f_o)\Phi|| \leq \delta||\Phi||$ for any Φ if $\mathcal{K}$ is closed, i.e. if the interval $\mathcal{K}$ is given by $|f - f_o| \leq \delta$;

2) $||(f(B) - f_o)\Phi|| < \delta||\Phi||$ for any $\Phi \neq 0$ if $\mathcal{K}$ is open, i.e. if the interval $\mathcal{K}$ is given by $|f - f_o| < \delta$.

To prove the main theorem we rely on the Weierstrass approximation theorem. Accordingly, to a given continuous function $f(\beta)$ there is a polynomial $p_\varepsilon(\beta)$ such that $|f(\beta) - p_\varepsilon(\beta)| \leq \varepsilon$ for $|\beta| \leq b$. Hence $|p_\varepsilon(\beta) - p_\delta(\beta)| \leq \varepsilon + \delta$ and the corollary yields the relation $||p_\varepsilon(B) - p_\delta(B)|| \leq \varepsilon + \delta$. Clearly, then the sequence of operators $p_\varepsilon(B)$ is a "uniform" Cauchy sequence as ε tends to zero. These operators, therefore, approach a limit operator, denoted by $f(B)$, such that $||p_\varepsilon(B) - f(B)|| \to 0$ as $\varepsilon \to 0$.

Note that the validity of Rule IV is implied by this construction.

Since the sums and products of approximating polynomials respectively tend to the sums and products of the approximated functions and since the product of two strongly converging operators converges strongly we conclude that also the rules I and II of the operational calculus are obeyed by the assignment of $f(B)$ to $f(\beta)$.

Furthermore, if $f(\beta) \geq 0$ for all β we have $p_\varepsilon(\beta) + \varepsilon \geq 0$, hence $p_\varepsilon B + \varepsilon \geq 0$, whence in the limit $f(B) \geq 0$.

Applying this result to the functions $f(\beta) - f_1$ and $f_2 - f(\beta)$, rule III follows for the closed interval $f_1 \leq f(\beta) \leq f_2$. To derive rule III for an open interval we need only consider the case in which $f(\beta) > 0$ for all β. Clearly, in that case there is an $a > 0$ such that $f(\beta) \geq a$ for all $|\beta| \leq b$. By virtue of rule IV we may change $f(\beta)$ outside of $|\beta| \leq b$ in such a way that $f(\beta) \geq a$ for all β. Then $f(B) \geq a > 0$ follows. It is thus seen that rule III holds also for an open interval $\mathscr{K}$.

It is not sufficient for our purposes to have operators assigned to continuous functions; we must set up such an assignment for piecewise continuous functions. Having done this we shall be able to establish "spectral resolution" of an Hermitean operator.

We first take step functions, i.e. characteristic functions of intervals $\mathscr{I}$ defined by

$$\eta_{\mathscr{I}}(\beta) = 1 \quad \text{for} \quad \beta \quad \text{in} \quad \mathscr{I}, = 0 \quad \text{for} \quad \beta \quad \text{outside} \quad \mathscr{I}.$$

We here regard also a single point as an interval, a closed one, of course.

If $\mathscr{I}$ is an open interval, $\beta_- < \beta < \beta_+$, we approximate $\eta_{\mathscr{I}}(\beta)$ from the inside by the piecewise linear functions

$$\begin{aligned} \eta^\delta_{\mathscr{I}}(\beta) &= \pm(\beta_\pm - \beta)/\delta \quad \text{for} \quad 0 \leq \pm(\beta_\pm - \beta) \leq \delta \\ &= 1 \quad \text{for} \quad \beta_- - \delta \leq \beta \leq \beta_+ + \delta, \\ &= 0 \quad \text{outside} \quad \mathscr{I}. \end{aligned}$$

Evidently, these functions increase monotonically as $\delta \to 0$.

If $\mathscr{I}$ is closed, $\beta_- \leq \beta \leq \beta_+$, we take

$$= 1 \quad \text{in} \quad \mathscr{I},$$

$$\eta^{\delta}(\beta) = 1 \pm (\beta_{\pm} - \beta)/\delta \quad \text{for} \quad \pm(\beta - \beta_{\pm}) \leq \delta,$$

$$= 0 \quad \text{for} \quad \pm(\beta-\beta_{\pm}) \geq \delta.$$

These functions decrease monotonically as $\delta \to 0$.

Since these functions $\eta^{\delta}_{\mathscr{I}}(s)$ are continuous we may conclude from rule III of the main theorem that the operators $\eta^{\delta}_{\mathscr{I}}(B)$ behave in the same way; i.e. the forms $(\Phi, \eta^{\delta}_{\mathscr{I}}(B)\ )$ increase monotonely if $\mathscr{I}$ is an open interval and decreases if $\mathscr{I}$ is closed.

We now may apply the Theorem 19.3 of monotone convergence formulated in Section 19, and conclude that the operators $\eta^{\delta}_{\mathscr{I}}(B)$ converge strongly to limit operators, which we shall denote by $\eta_{\mathscr{I}}(B)$.

To formulate a counterpart of rule I for the operators $\eta_{\mathscr{I}}(B)$ we assume that the interval $\mathscr{I}$ is the union of three other adjacent intervals,

$$\mathscr{I} = \mathscr{I}_{-} \cup \mathscr{I}_{o} \cup \mathscr{I}_{+},$$

where either $\mathscr{I}_{+}$ and $\mathscr{I}_{-}$ are open and $\mathscr{I}_{o}$ closed, or $\mathscr{I}_{+}$ and $\mathscr{I}_{-}$ are closed and $\mathscr{I}_{o}$ open. In either case we have

$$I_{\eta}: \qquad \eta_{\mathscr{I}_{-}}(B) + \eta_{\mathscr{I}_{o}}(B) + \eta_{\mathscr{I}_{+}}(B) = \eta_{\mathscr{I}}(B).$$

To verify that this is so we need only observe that the piecewise linear functions that approximate the three step functions can be so chosen that in their sum the linear sections interior to $\mathscr{I}$ cancel away, so that this sum is an approximating function for the total interval $\mathscr{I}$.

On occasion we shall express a closed interval as the union of the open interior and the two endpoints, regarded as closed intervals. Formula I_{η} then allows us to express the operator $\eta_{\mathscr{I}}(B)$ for such a closed interval as the sum of the operators $\eta(B)$ associated with these three parts of it. Also we observe that if an endpoint of an interval lies outside of $|\beta| \leq b$ it does not matter where it lies in $\beta > b$ or $\beta < -b$, by virtue of rule IV. We, therefore, may just as well allow an interval to have an endpoint at $\beta = \infty$ or at $\beta = -\infty$.

In deriving rule II for the operators $\eta_{\mathscr{I}}(B)$ and related facts we shall rely on a lemma concerning the operator $f(B)\eta_{\mathscr{I}}(B)$ formed with a continuous function $f(\beta)$ of β. Clearly, this operator is defined as the limit of the operators $f(B)\eta^{\delta}_{\mathscr{I}}(B)$ which exists since $\eta^{\delta}_{\mathscr{I}}(B)$ converges strongly to $\eta_{\mathscr{I}}(B)$ if $\delta \to 0$.

<u>Lemma 1</u>. Suppose the continuous function $f(\beta)$ is non-negative in an interval $\mathscr{I}$ and has the absolute bound κ there, so that

$$0 \leq f(\beta)\eta_{\mathscr{I}}(\beta) \leq \kappa \quad \text{for all} \quad \beta;$$

then for all vectors Φ the inequality

$$(\Phi, f(B)\eta_{\mathscr{I}}(B)\Phi) \leq \kappa(\Phi,\Phi)$$

holds.

To prove this statement we choose, for a given $\varepsilon > 0$, the approximating function $\eta^{\delta}_{\mathscr{I}}(\beta)$ in such a way that

$$-\varepsilon \leq f(\beta)\eta^{\delta}_{\mathscr{I}}(\beta) \leq \kappa + \varepsilon,$$

which is possible by virtue of the continuity of f. From rule III we then conclude that the inequality

$$-\varepsilon(\Phi,\Phi) \leq (\Phi, f(B)\eta^{\delta}_{\mathscr{I}}(B)\Phi) \leq (\kappa+\varepsilon)(\Phi,\Phi)$$

holds. Since $\eta^{\delta}_{\mathcal{I}}(B)$ converges strongly to $\eta_{\mathcal{I}}(B)$ and ε is arbitrary, the statement follows.

Taking $\kappa = 0$ we are immediately led to the

Corollary to Lemma 1. Suppose the continuous function $f(\beta)$ vanishes in the interval $\mathcal{I}$; then

$$f(B)\eta_{\mathcal{I}}(B) = 0.$$

From Lemma 1 we can derive an important second lemma which refers to a monotone sequence of intervals $\mathcal{I}_{\sigma}$ with empty intersection; i.e. $\sigma < \tau$ implies $\mathcal{I}_{\sigma} \subset \mathcal{I}_{\tau}$ and $\bigcap_{\sigma} \mathcal{I}_{\sigma} = \emptyset$. The statement is that the operators $\eta_{\sigma}(B) = \eta_{\mathcal{I}_{\sigma}}(B)$ tend to zero for such a sequence. It is sufficient for our purpose (and actually no restriction) to consider a special case.

Lemma 2. Let $\mathcal{I}_{\sigma}$ be the open interval $0 < \beta < \sigma$. Then the operators $\eta_{\sigma}(B)$ tend to zero.

Proof. If σ tends to 0 monotonically the functions $\eta_{\sigma}(\beta)$ form a monotone sequence; hence the operators $\eta_{\sigma}(B)$ tend strongly to a limit which we denote by Q. Let $\eta^{\delta}_{\tau}(\beta)$ be an approximating piecewise linear function for $\eta_{\tau}(\beta)$. Then we can, for every $\varepsilon > 0$, find a $\sigma \leq \tau$ such that

$$\eta_{\sigma}(\beta)\eta^{\delta}_{\tau}(\beta) \leq \varepsilon.$$

By the lemma we have, for every Φ,

$$(\Phi, \eta_{\sigma}(B)\eta^{\delta}_{\tau}(B)\Phi) \leq \varepsilon(\Phi, \Phi).$$

Letting successively σ, δ, τ, and ε tend to zero we find successively,

$$(\Phi, Q\eta_\tau^\delta(B)\Phi) \leq \varepsilon(\Phi,\Phi), \quad (\Phi, Q\eta(B)\Phi) \leq \varepsilon(\Phi,\Phi),$$
$$(\Phi, Q^2\Phi) \qquad \leq \varepsilon(\Phi,\Phi), \quad (\Phi, Q^2\Phi) = 0.$$

Clearly, Q is self-adjoint, since the $\eta_\sigma(B)$ are; hence $||Q\Phi||^2 = (\Phi, Q^2\Phi) = 0$ and therefore $Q = 0$. Thus the statement of the theorem is proved.

An immediate consequence of this theorem is the

Corollary to Lemma 2. Let $\mathscr{I}$ be an open (or closed) interval and $\mathscr{I}_\sigma$ be a monotonely increasing (or decreasing) sequence of intervals covering all of $\mathscr{I}$, so that $\cup_\sigma \mathscr{I}_\sigma = \mathscr{I}$. Then

$$\eta_{\mathscr{I}_\sigma}(B) \to \eta_{\mathscr{I}}(B).$$

Without restriction one may assume $\mathscr{I}_\sigma$ to be closed if $\mathscr{I}$ is open; then one needs only apply the theorem to the two shrinking sequences of open intervals left over from $\mathscr{I}$ after $\mathscr{I}_\sigma$ has been removed. The statement for the case that $\mathscr{I}$ is closed then follows by going over to the complement of $\mathscr{I}$.

We are now ready to prove rule II for step functions. This rule assumes a special form since the product of two step functions is again a step function. Specifically, if $\mathscr{I}$ is the intersection of two intervals $\mathscr{I}_1$, $\mathscr{I}_2$,

$$\mathscr{I}_1 \cap \mathscr{I}_2 = \mathscr{I},$$

its step function is the product of those of the two intervals:

$$\eta_{\mathscr{I}_1}(\beta)\eta_{\mathscr{I}_2}(\beta) = \eta_{\mathscr{I}}(\beta).$$

Accordingly, we expect rule II for step functions to be

II_η: $$\mathscr{I}_1 \cap \mathscr{I}_2 = \mathscr{I}$$

implies

$$\eta_{\mathscr{I}_1}(B)\eta_{\mathscr{I}_2}(B) = \eta_{\mathscr{I}}(B).$$

In proving that this rule is valid we shall employ piecewise linear approximating functions $\eta^{\delta_1}_{\mathscr{I}_1}(\beta)$ and $\eta^{\delta_2}_{\mathscr{I}_2}(\beta)$ and use the relation

$$\eta^{\delta_1}_{\mathscr{I}_1}(B)\eta^{\delta}_{\mathscr{I}_2}(B) \to \eta_{\mathscr{I}_1}(B)\eta_{\mathscr{I}_2}(B)$$

if $\delta_1 \to 0$ and $\delta_2 \to 0$ in any manner. Actually, it will turn out to be sufficient to prove rule II_η in cases in which the intersection is empty, $\mathscr{I}_1 \cap \mathscr{I}_2 = \emptyset$. In such cases we need only show that

$$\eta^{\delta_1}_{\mathscr{I}_1}(B)\eta^{\delta_2}_{\mathscr{I}_2}(B) \to 0 \qquad \text{as} \qquad \delta_1 \to 0, \quad \delta_2 \to 0.$$

There are four such cases. In case both $\mathscr{I}_1$ and $\mathscr{I}_2$ are open with empty intersection, the product of the approximating functions is zero; hence the rule holds with $\eta_{\mathscr{I}}(B) = 0$. The same argument applies when the two intervals are at a positive distance from each other. If one of these intervals is closed one must, of course, choose the linear parts of the associated approximating functions sufficiently steep.

In case one interval, $\mathscr{I}_1$, is open and the other, $\mathscr{I}_2$, a boundary point β_2 of $\mathscr{I}_1$ we apply the corollary of Lemma 1 to the interval $\mathscr{I}_2$ with $f(\beta) = \eta_1^\delta(\beta)$ being an approximation to $\eta_1(\beta)$. Since evidently $f(\beta_2) = 0$ we have $\eta_1^\delta(B)\eta_2^\delta(B) = 0$ and hence $\eta_1(B)\eta_2(B) = 0$.

To handle the general case one should observe that any two intervals can be referred to a common subdivision and be written as the sum of open intervals and vertices of this subdivision. The products of the two step functions are then sums of products, either of the four special types discussed first or of the products of the form $\eta_{\mathscr{I}}^2$ with $\mathscr{I}$ being open or a point. To handle these two cases one need only write

$$\eta_{\mathscr{I}}^2 = \eta - \eta_{\mathscr{I}}\eta_{\mathscr{I}*} ,$$

when $\mathscr{I}^*$ is the complement of $\mathscr{I}$, which evidently can be written as the union of intervals either away from $\mathscr{I}$ or adjacent to $\mathscr{I}$, but not intersecting $\mathscr{I}$. That is to say, the product $\eta_{\mathscr{I}}(B)\eta_{\mathscr{I}*}(B)$ consists of the sum of products of the four types considered first and hence is zero. Thus the relation

$$\eta_{\mathscr{I}}^2 (B) = \eta_{\mathscr{I}}(B)$$

results. Rule II_η has thus been established.

Relation $\eta_{\mathscr{I}}^2(B) = \eta_{\mathscr{I}}(B)$ implied by rule II_η shows that the operators

$$P_{\mathscr{I}} = \eta_{\mathscr{I}}(B)$$

are projectors. Moreover, rule II_η shows that the projectors of foreign intervals annihilate each other,

$$P_{\mathscr{I}_1} P_{\mathscr{I}_2} = 0 \quad \text{if} \quad \mathscr{I}_1 \cap \mathscr{I}_2 = \emptyset.$$

The projector $P_{|\beta| \le b}$ assigned to the interval $|\beta| \le b$ is the identity,

$$P_{|\beta| \le b} = 1,$$

as follows from rule IV and the corollary to Lemma 1.

The assignment of the projectors $P_{\mathscr{I}} = \eta_{\mathscr{I}}(B)$ to the intervals $\mathscr{I}$ constitutes the spectral resolution of the operator B. The range of the projector $P_{\mathscr{I}}$ will be called the eigenspace of the operator B for the interval $\mathscr{I}$ and the vectors $P_{\mathscr{I}}\Phi$ in this range will be called eigenvectors of B for the interval $\mathscr{I}$. This terminology is justified by the fact that these vectors are eigenvectors in the original sense in case the interval consists just of one point. That this is so will be implied by the subsequent considerations.

For eigenvectors of an interval we can formulate a rule analogous to rule III.

$\text{III}_{\mathscr{I}}$: For all eigenvectors $\Phi = P_{\mathscr{I}}\Phi \neq 0$ of the interval $\mathscr{I}$ the mean value

$$(\Phi, f(B)\Phi)/(\Phi,\Phi) \quad \text{lies in } \mathscr{K}$$

if the continuous function $f(\beta)$ takes its values in the interval $\mathscr{K}$ for β in $\mathscr{I}$.

We here assume, without serious restriction, that $\mathscr{K}$ is closed if $\mathscr{I}$ is closed and that $\mathscr{K}$ is open if $\mathscr{I}$ is open.

In case $\mathscr{I}$ and $\mathscr{K}$ are closed, the statement is an immediate consequence of the lemma ; one need only replace $(P_{\mathscr{I}}\Phi, f(B)P_{\mathscr{I}}\Phi)$ by $(P\Phi, f(B)\eta_{\mathscr{I}}(B)\Phi)$.

In case the interval $\mathscr{K}$ is open, given by $f_- < f < f_+$, we first observe that

$$(\Phi, f(B)\Phi) \geq f_-(\Phi,\Phi)$$

for eigenvectors Φ of $\mathscr{I}$; for, these eigenvectors are also eigenvectors of the closure of $\mathscr{I}$ so that the statement for closed intervals can be employed. Now we take any eigenvector Φ of $\mathscr{I}$ for which

$$(\Phi, f(B)\Phi) = f_-(\Phi,\Phi)$$

and prove that $\Phi = 0$.

To this end we take any closed interval $\mathscr{I}_\sigma$ in the interior of $\mathscr{I}$ and set $\eta_{\mathscr{I}_\sigma}(B) = Q_\sigma$. Clearly, the eigenvectors of Q_σ are eigenvectors of $P_{\mathscr{I}}$ and, since $f(B) - f_-$ is nonnegative for these eigenvectors Φ we have

$$0 \leq (Q_\sigma\Phi, (f(B) - f_-)Q_\sigma\Phi) \leq (\Phi, (f(B) - f_-)\Phi).$$

Hence

$$(Q_\sigma\Phi, (f(B) - f_-)Q_\sigma\Phi) = 0.$$

Since the values of $f(\beta) - f_-$ are positive by assumption there is a positive lower bound a_σ for this function in the interval $\mathscr{I}_\sigma$; therefore

$$(Q_\sigma\Phi,(f(B) - f_-)Q_\sigma\Phi) \geq a_\sigma(Q_\sigma\Phi,Q_\sigma\Phi).$$

It follows that $Q_\sigma\Phi$ is zero. Now, if one lets the intervals $\mathscr{I}_\sigma$ increase so as to cover all of $\mathscr{I}$, the operators Q_σ tend to $P_\mathscr{I}$ by virtue of the corollary to Lemma 2 proved earlier. It follows that $P_\mathscr{I}\Phi = 0$. But, $P_\mathscr{I}\Phi = \Phi$, and hence $\Phi = 0$. Thus we have proved that

$$(\Phi,f(B)\Phi) > f_-(\Phi,\Phi)$$

for all $\Phi = P\Phi$. Similarly one proves

$$(\Phi,f(B)\Phi) < f_+(\Phi,\Phi)$$

and thus the statement.

The following converse of rule $III_\mathscr{I}$ can be proved.

$III^*_\mathscr{I}$: Suppose the function $f(\beta)$ has its values in the interval $\mathscr{K}$ for β in $\mathscr{I}$ and outside of $\mathscr{K}$ for β outside of $\mathscr{I}$. Suppose moreover that Φ is a vector for which

$$(\Phi,f(B)\Phi)/(\Phi,\Phi) \quad \text{lies in} \quad \mathscr{K}.$$

Then Φ is an eigenvector of $\mathscr{I}$.

To prove it one need only apply the second part of rule $III_\mathscr{I}$ to the two open sets complementary to the closed set $\mathscr{I}$ to obtain the converse of the first part, and conversely apply the first part of the two closed sets complementary to the open set $\mathscr{I}$ to obtain the converse of the second part.

We shall not give the details.

The first part of rule $III_\mathscr{I}$, when applied to an interval $\mathscr{I}$

that consists just of one point β_o, yields the statement that, for all eigenvectors of this point interval,

$$(\Phi,(f(B) - f(\beta_o))\Phi) = 0.$$

It follows that the operator $(f(B) - f(\beta_o))P_{\mathcal{I}}$ is zero. That is to say, all eigenvectors $\Phi = P_{\mathcal{I}}\Phi$ of this point satisfy the relation

$$f(B)\Phi = f(\beta_o)\Phi;$$

in particular,

$$B\Phi = \beta_o\Phi.$$

These eigenvectors are thus seen to be eigenvectors in the usual sense.

The converse is also true as seen by applying rule $III^*_{\mathcal{I}}$ to the function $f(\beta) = \beta$.

Finally we introduce the class of "piecewise continuous" functions as the functions which can be written in the form

$$F_\beta: \qquad f(\beta) = \sum_{\mathcal{I}} f_{\mathcal{I}}(\beta)\eta_{\mathcal{I}}(\beta)$$

where each function $f_{\mathcal{I}}(\beta)$ is continuous in $|\beta| \leq b$. We can now assign to such a function an operator

$$F_B: \qquad f(B) = \sum_{\mathcal{I}} f_{\mathcal{I}}(B)\eta_{\mathcal{I}}(B).$$

From the corollary to Lemma 1 we infer that this operator is independent of the choice of the function $f_{\mathcal{I}}(\beta)$ employed to express it. From rule $I_{\mathcal{I}}$ we infer that the operator is independent of the choice of the partition to which the intervals $\mathcal{J}$ belong. Rule I then clearly holds for our piecewise continuous functions. From rule $II_{\mathcal{J}}$ and $III_{\mathcal{J}}$ we conclude that also rules II and III hold for these functions. That is to say the assignment of $f(B)$ to $f(\beta)$ obeys the first three rules of operational calculus.

Accordingly we may formulate the

Corollary to the Main Theorem. To every ordinary piecewise continuous function $f(\beta)$ of the form F_β there can be assigned, through F_B, an operator $f(B)$ such that the four rules I, II, III, IV of operational calculus are observed.

22. Spectral Representation

Having established a functional calculus for a bounded Hermitean operator by operators $f(B)$ to ordinary piecewise continuous functions $f(\beta)$ we have established the spectral resolution of the operator B, which is given by assigning projectors $\eta_{\mathcal{J}}(B)$ to intervals $\mathcal{J}$ of a partition. We now proceed to establish a spectral representation.

To this end we first select any vector Ω from the Hilbert space $\mathfrak{H}$ and form the subspace of $\mathfrak{H}$ which consists of all vectors of the form

$$\Phi = h(B)\Omega = f(B)\Omega + ig(\beta)\Omega.$$

where

$$h(\beta) = f(\beta) + ig(\beta)$$

runs over all piecewise continuous functions.

This subspace will be denoted by $\mathfrak{G}'(\Omega)$; its closure will be denoted by $\mathfrak{G}(\Omega)$ and called the space "generated" by Ω.

With this space we shall associate a measure function $r(s)$. To this end we consider the open interval $\mathscr{I}_\beta$ with the upper end point β while the lower end point is any number $< -b$. The step function of β' associated with this interval will be denoted by $\eta_\beta^-(\beta')$. The step function of β' of the open interval from β to any point $< b$ will be denoted by $1 - \eta_\beta^+(\beta')$. In other words we have

$$\eta_\beta^-(\beta') = 1 \quad \text{for} \quad \beta' < \beta, \; = 0 \quad \text{for} \quad \beta' \geq \beta,$$

$$\eta_\beta^+(\beta') = 1 \quad \text{for} \quad \beta' \leq \beta, \; = 0 \quad \text{for} \quad \beta' > \beta.$$

Now we introduce the functions

$$r^\pm(\beta) = ||\eta_\beta^\pm(B)\Omega||^2 = (\Omega, \eta_\beta^\pm(B)\Omega).$$

Evidently, these functions are monotonically increasing. We remark that <u>both</u> $r^+(\beta)$ <u>and</u> $r^-(\beta)$ <u>tend to</u> $r^+(\beta_o)$ <u>when</u> β <u>tends to</u> β_o <u>from above, and to</u> $r^-(\beta_o)$ <u>when</u> β <u>tends to</u> β_o <u>from below</u>. This follows from the corollary to Lemma 2. For, the step function of β' of the open interval $\beta_o < \beta' < \beta$ is $\eta_\beta^-(\beta') - \eta_{\beta_o}(\beta')$. Therefore, $\eta_\beta^-(B) - \eta_\beta^+(B) \to 0$ and hence $r^-(\beta) - r^+(\beta_o) \to 0$ as $\beta \downarrow \beta_o$. Applying this to any $\beta_1 > \beta$ instead of β, and observing $r^+(\beta) \leq r^-(\beta_1)$ one finds $r^+(\beta) - r^+(\beta_o) \to 0$. Similarly, one derives the statement for $\beta \uparrow \beta_o$.

The remark made shows that the pair $r^\pm(\beta)$ could indeed be used as a "measure function pair" as introduced in Section 8, Chapter II.

We now consider a partition $\mathscr{P}$ of the interval $|\beta| < b$ into open intervals and points. In agreement with the procedure in Section 8 we set

$$\mathscr{I}_\sigma:\ \beta_{\sigma-1} < \beta < \beta_{\sigma+1}, \quad \sigma \text{ even, open, interval,}$$

$$\mathscr{I}_\sigma:\ \beta = \beta_\sigma, \quad \sigma \text{ odd, point.}$$

$$\Delta_\sigma r = r^-(\beta_{\sigma+1}) - r^+(\beta_{\sigma-1}) \quad \text{for} \quad \sigma \quad \text{even,}$$

$$\Delta_\sigma r = r^+(\beta_\sigma) - r^-(\beta_\sigma) \quad \text{for} \quad \sigma \quad \text{odd.}$$

Clearly,

$$\Delta_\sigma r = ||\eta_{\mathscr{I}_\sigma}(B)\Omega||^2 = (\Omega, \eta_{\mathscr{I}_\sigma}(B)\Omega).$$

With complex constants h_σ we form the piecewise constant function

$$k(\beta) = \sum_\sigma h_\sigma \eta_{\mathscr{I}_\sigma}(\beta)$$

and the operator

$$h(B) = \sum_\sigma h_\sigma \eta_{\mathscr{I}_\sigma}(B).$$

By virtue of

$$\begin{aligned} \eta_{\mathscr{I}_\tau}(\beta)\eta_{\mathscr{I}_\sigma}(\beta) &= \eta_{\mathscr{I}_\sigma}(\beta) \quad &\text{for} \quad \tau = \sigma \\ &= 0 \quad &\text{for} \quad \tau \neq \sigma \end{aligned}$$

we have

$$||h(B)\Omega||^2 = \sum_\sigma |h_\sigma|^2 \Delta_\sigma r.$$

Now, the sum here can be interpreted as the integral of $|h(\beta)|^2$ with respect to the measure pair $r^{\pm}(s)$. In other words,

$$||h(B)\Omega||^2 = \int |h(\beta)|^2 dr(\beta).$$

Let now $h(\beta)$ be a function in $\mathfrak{C}'$, i.e. a piecewise continuous function; since it can be approximated uniformly by piecewise constant functions $h^{\lambda}(\beta)$, we clearly have

$$||(h(B) - h^{\lambda}(B))\Omega||^2 \to 0 \quad \text{as well as} \quad \int |h(\beta) - h^{\lambda}(\beta)|^2 dr(s) \to 0$$

as $\lambda \to \infty$. Evidently then

$$||h(B)||^2 = \int |h(\beta)|^2 dr(\beta)$$

and

$$(h(B)\Omega, h'(B)\Omega) = \int \overline{h(\beta)} h'(\beta) dr(\beta).$$

for any function in this class $\mathfrak{C}'$.

We can go one step further. We may introduce the closure $\hat{\mathfrak{C}}_r$ of the space $\mathfrak{C}'$ of functions with respect to the unit form $\int |h(\beta)|^2 dr(\beta)$ by adjoining ideal functions $h(\beta)$. Approximating $\int |h(\beta) - h^{\lambda}(\beta)|^2 dr(\beta) \to 0$. Now, the sequence $h^{\lambda}(\beta)$ forms a Cauchy sequence and hence the vectors $h^{\lambda}(B)\Omega$ form a Cauchy sequence; they converge to a vector Φ such that $||\Phi - h^{\lambda}(B)\Omega|| \to 0$. Clearly, then the ideal functions $h(\beta)$ correspond to vectors in the space $\mathfrak{C}(\Omega)$ generated by Ω. The converse, of course, holds true just as well.

Thus we have established a one-to-one correspondence of the

vectors in $\mathfrak{C}(\Omega)$ and the functions in the closed function space $\hat{\mathfrak{C}}_r$,

$$\Phi \Leftrightarrow h(\beta)$$

such that

$$(\Phi,\Phi) = \int |h(\beta)|^2 dr(\beta) .$$

Of course, we must set $h(\beta) = 0$ whenever $\int |h(\beta)|^2 dr(\beta) = 0$. It is also evident that the vector $B\Phi$ corresponds to the function $\beta h(\beta)$ since for functions in $\mathfrak{C}'$, the relation $\beta h(\beta) = h_1(\beta)$ leads to the relation $Bh(B)\Omega = h_1(B)\Omega$ by rule II of the functional calculus. In other words

$$B\Phi \Leftrightarrow \beta h(\beta) .$$

Thus we have achieved a spectral representation of the operator B in the subspace $\mathfrak{C}(\Omega)$ of the Hilbert space $\mathfrak{H}$.

Symbolically, we may express this relationship by the formula

$$\Phi = h(B)\Omega$$

for all ideal functions in $\hat{\mathfrak{C}}_r$.

From here it is only one step to the full spectral representation of $\mathfrak{H}$.

For simplicity we assume the space $\mathfrak{H}$ to be of (at most) countable dimension. Then there is certainly a sequence of vectors $\Omega_1, \Omega_2, \ldots$ in $\mathfrak{H}$ which spans the space $\mathfrak{H}$ densely. That is, to every Φ and $\varepsilon > 0$ there is an $n = n_\varepsilon$ and a linear combination $\Phi^{(n)}$ of $\Omega_1, \ldots, \Omega_n$ such that

$$||\Phi-\Phi^{(n)}|| \leq \varepsilon.$$

The space of linear combinations of $\Omega_1,\ldots,\Omega_n$ will be denoted by

$$\mathfrak{T}_n = \mathfrak{T}(\Omega_1,\ldots,\Omega_n)$$

so that we may say that $\Phi^{(n)}$ is in $\mathfrak{T}_n$.

Now set $\Omega^{(1)} = \Omega_1$ and let P_1 be the projector which projects into the space $\mathfrak{G}^{(1)} = \mathfrak{G}(\Omega^{(1)})$ generated by Ω_1 and let successively

$$\Omega^{(2)} = (1-P_1)\Omega_2, \text{ and } P_2 \text{ be the projector into}$$

$$\mathfrak{G}_n^{(2)} = \mathfrak{G}(\Omega^{(2)}),$$

$$\Omega^{(n)} = (1-P_1-\cdots-P_{n-1})\Omega_n, \text{ and } P_n \text{ be the projector into}$$

$$\mathfrak{G}_n = \mathfrak{G}(\Omega^{(n)}).$$

Then, we maintain, $\mathfrak{G}^{(n)}$ is perpendicular to $\mathfrak{G}^{(1)},\ldots,\mathfrak{G}^{(n-1)}$ or, in other words, $P_nP_1 = P_nP_2 \cdots = P_nP_{n-1} = 0$, for all n.

To show this assume the statement proved up to $n-1$, so that $(P_1 + \cdots + P_{n-1})\Psi^{(r)} = \Psi^{(r)}$ for every $\Psi^{(r)}$ in $\mathfrak{G}^r$. Now, let $\Phi_n = h_n(B)\Omega^{(n)}$ be in $\mathfrak{G}^{(n)}$ and $\Phi_r = h_r(B)\Omega^{(r)}$ be in $\mathfrak{G}^{(r)}$, where $h_n(\beta)$ and $h_r(\beta)$ are continuous functions. Then $\Psi^{(r)} = \bar{h}_n(B)h_r(B)\Omega^{(r)}$ is also in $\mathfrak{G}^{(r)}$ and

$$(\Phi_n,\Phi_r) = (h_n(B)\Omega^{(n)}, h_r(B)\Omega^{(r)}) = (\Omega^{(n)}, \Psi^{(r)})$$
$$= (\Omega_n, (1-P_1 - \cdots - P_{n-1})\Psi^{(r)}) = 0.$$

Hence the statement is true for n.

Next we maintain that the space $\mathfrak{G}_n$ is contained in $\mathfrak{G}^{(1)} \oplus \cdots \oplus \mathfrak{G}^{(n)}$. To show this we set

$$\Omega_n = P_1\Omega_n + \cdots + P_{n-1}\Omega_n + (1-P_1 - \cdots - P_{n-1})\Omega_n.$$

The first $n - 1$ vectors on the right hand side are in $\mathfrak{G}^{(1)} \oplus \cdots \oplus \mathfrak{G}^{(n-1)}$ by definition of $P_1,\ldots,P_{n-1}$; the last vector is in $\mathfrak{G}^{(n)}$ by definition of this space. Assuming the statement has been proved up to $n - 1$ it follows for n.

Now let Φ be any vector in $\mathfrak{H}$. We maintain that $(P_1 + \cdots + P_n)\Phi$ tends to Φ. In fact, for a given $\varepsilon > 0$ take the vector $\Phi^{(n)}$ in $\mathfrak{T}_n$ for which $||\Phi-\Phi^{(n)}|| \leq \varepsilon$. Since we now know that $\Phi^{(n)}$, being in $\mathfrak{T}_n$, is in $\mathfrak{G}^{(1)} \oplus \cdots \oplus \mathfrak{G}^{(n)}$ we can use the minimum property of the projection $P_1 + \cdots + P_n$ into this space. We find

$$||\Phi - (P_1 + \cdots + P_n)\Phi|| \leq ||\Phi-\Phi^{(n)}|| \leq \varepsilon .$$

Then the statement follows.

This statement implies that every vector Φ can be written in the form

$$\Phi = P_1\Phi + P_2\Phi + \cdots$$

as a series of vectors in $\mathfrak{G}^{(1)}, \mathfrak{G}^{(2)},\ldots$ so that

$$||\Phi||^2 = ||P_1\Phi||^2 + ||P_2\Phi||^2 + \cdots .$$

Now each vector in $\mathfrak{G}^{(n)}$ can be written in the form $h_n(B)\Omega^{(n)}$ with an appropriate (ideal) function $h_n(\beta)$. In other words, the vector Φ can be represented by a sequence of functions

$$\Phi \Leftrightarrow \{h_1(\beta),\ h_2(\beta),\dots\}$$

in the spaces $\hat{\mathfrak{C}}_{r_1}, \hat{\mathfrak{C}}_{r_2},\dots$ such that

$$(\Phi,\Phi) = \int |h_1(\beta)|^2 dr_1(\beta) + \int |h_2(\beta)|^2 dr_2(\beta) + \cdots .$$

At the same time $B\Phi$ is represented as

$$B\Phi \Leftrightarrow \{\beta h_1(\beta),\ \beta h_2(\beta),\dots\}.$$

Thus we have established one of our major aims: we have established a spectral representation of any bounded Hermitean operator.

Of course, we do not claim that this representation, which is of the direct multiple type, is the only one - or the most suitable one. In particular, it could happen that some of the measure functions are identically zero so that the corresponding terms drop out.

We also should mention - as Hellinger in 1909 has shown - that a "minimal" spectral representation can be established, i.e. one in which $\Delta_{\mathcal{J}} r_m = 0$ for some interval $\mathcal{J}$ always implies $\Delta_{\mathcal{J}} r_\ell = 0$ for $\ell > m$. We do not intend to discuss the proof of this fact.

Finally, we should mention that we could eliminate the assumption that the Hilbert space $\mathfrak{H}$ should have a countable dimension. To handle spectral representation in a non-countable Hilbert space one may employ a well ordered set $\{\Omega_m\}$ of vectors which span $\mathfrak{H}$ densely. All arguments given can then be carried over; we do not want to give details.

23. Normal and Unitary Operators

Suppose B and C are two bounded Hermitean operators which commute:

$$BC = CB.$$

Then powers, polynomials, and hence piecewise continuous functions of B commute with such functions of C. We may plot the eigenvalues β and γ of B and C in a (β,γ)-plane and introduce as common spectral resolution of the pair B,C the projectors $\eta_{\mathcal{J}}(B), \eta_{\mathcal{J}}(C)$ which correspond to the step function $\eta_{\mathcal{J}}(\beta), \eta_{\mathcal{J}}(\gamma)$ of the product cell $\mathcal{J} \times \mathcal{J}$. It is also clear that to any bounded piecewise continuous function $f(\beta,\gamma)$ an operator $f(B,C)$ can be assigned obeying the rules of the operational calculus.

With the aid of two such commuting operators we may form the operator

$$B + iC$$

which, unless $C = 0$, is not Hermitean. Any such operator is called "normal". The eigenvalues $\beta + i\gamma$ of $B + iC$ are complex numbers. The common spectral resolution of B and C may then be regarded as the spectral resolutions of $B + iC$. Also, it is clear that to every (complex-valued) bounded piecewise continuous function $f(\beta+i\gamma)$ of $\beta + i\gamma$ an operator $f(B+iC)$ may be assigned obeying the rules of operational calculus.

In particular, if Φ is an eigen-function of the product cell $\mathcal{J} \times \mathcal{J}$ the value of the ratio

$$(\Phi, (B+iC)\Phi)/(\Phi,\Phi)$$

lies in this cell, as follows from Rule $III_{\mathcal{J}}$ in Section 21.

If the normal operator $U = B + iC$ together with the adjoint $U^* = B - iC$ satisfies the condition

$$U^*U = 1$$

it is called "unitary". Since U and U^* commute, i.e. $UU^* = U^*U$, we have also $UU^* = 1$, so that $U^* = U^{-1}$. Clearly, a unitary operator is "norm preserving"

$$||U\Phi|| = ||\Phi||.$$

Let $\mathscr{I} \times \mathscr{J}$ be a cell in which $0 \leq \beta_1 \leq |\beta| \leq \beta_2$ and $0 \leq \gamma_1 \leq |\gamma| \leq \gamma_2$ with either $\beta_1^2 + \gamma_1^2 > 1$ or $\beta_2^2 + \gamma_2^2 > 1$. Let $\Phi \neq 0$ be an eigenvector of this cell. From Rule $III_{\mathscr{I}}$ of Section 21 applied to B^2 and C^2 we conclude that the ratio

$$(\Phi,(B^2 + C^2)\Phi)/(\Phi,\Phi)$$

is either greater or less than 1, which contradicts the condition $B^2 + C^2 = 1$. Hence there is no such eigenvector $\Phi \neq 0$. It then follows that the eigenspaces of any product cell vanish if this cell lies outside of the unit circle. We express this fact by saying that <u>the spectrum of a unitary operator lies on the unit circle</u>.

As a consequence of this fact we note down the following

<u>Lemma</u>. Suppose the piecewise continuous function $f(\beta,\gamma)$ vanishes on the unit circle; then $f(B,C) = 0$.

For, such a function may be written in the form

$$f(\beta,\gamma) = \sum_{\mathscr{I},\mathscr{J}} f_{\mathscr{I},\mathscr{J}}(\beta,\gamma)\eta_{\mathscr{I}}(\beta)\eta_{\mathscr{J}}(\gamma)$$

with continuous functions $f_{\mathscr{I},\mathscr{J}}(\beta,\gamma)$. By subdivision one can achieve that $|f_{\mathscr{I},\mathscr{J}}(\beta,\gamma)| \leq \varepsilon$ in those cells that intersect the unit circle. The contributions from the product cells $\mathscr{I} \times \mathscr{J}$ that lie outside of the unit circle may evidently be omitted. Consequently we have $||f(B,C)|| \leq \varepsilon$, hence $f(B,C) = 0$ since ε is arbitrary.

We now let $\theta(\beta,\gamma)$ be any real-valued piecewise continuous function of β and γ for which

$$e^{i\theta(\beta,\gamma)} = \beta + i\gamma$$

on $\beta^2 + \gamma^2 = 1$. To be sure such a function can be constructed; one need only take a function whose values on the unit circle agree with the polar angle θ restricted by $-\pi < \theta \leq \pi$.

From the lemma we then conclude that the relation

$$e^{i\theta(B,C)} = B + iC = U$$

holds. Now, the operator $\Theta = \theta(B,C)$ is self-adjoint since $\theta(\beta,\gamma)$ is real-valued. Thus, we have established the important fact that <u>every unitary operator</u> U <u>can be written in the form</u>

$$U = e^{i\Theta}$$

with the aid of a bounded self-adjoint operator Θ with spectrum in $-\pi < \theta \leq \pi$.

CHAPTER V

OPERATORS WITH DISCRETE SPECTRA

24. Operators with Partly Discrete Spectra

There are various classes of operators whose spectra have significant special properties. In this chapter we shall discuss operators whose spectra are "discrete" or "partly discrete".

A discrete spectrum is a pure point spectrum; but the term "discrete" is to imply more, namely, that each point eigenvalue has a finite multiplicity and that there is only a finite number of eigenvalues in each interval. This requirement is rather severe. We shall require somewhat less by allowing the eigenvalues to accumulate at zero, and by allowing zero itself to be an eigenvalue of infinite multiplicity. We then say the spectrum is "discrete away from zero", or simply "essentially discrete".

We may describe the property of discreteness in an interval $\mathcal{J}$ as a property of the eigenspace associated with this interval, without mentioning point eigenvalues explicitly. In doing this we assume the operator - denoted by K - to be bounded and Hermitian, so that we can refer to its spectral resolution, i.e. to the projectors $\eta_{\mathcal{J}}(K)$ and the eigenspaces $\mathfrak{E}_{\mathcal{J}}$ associated with intervals $\mathcal{J}$ of the κ-axis.

Using these notions, we say that the spectrum of the operator K is discrete in an interval $\mathcal{J}$ if the eigenspace $\mathfrak{E}_{\mathcal{J}}$ has a finite dimension. In particular we shall consider for any positive λ the intervals $\mathcal{J}_\lambda$: $\lambda < \kappa < \infty$ (or $\lambda < \kappa \leq ||K||$) and $\mathcal{J}_{-\lambda}$: $-\infty < \kappa < -\lambda$ (or $-||K|| \leq \kappa < \lambda$) and denote by $\mathfrak{E}_\lambda$ and $\mathfrak{E}_{-\lambda}$ the associated eigenspaces. Then we shall say: <u>the spectrum of the operator</u> K <u>is discrete above</u> $\lambda > 0$ (<u>or below</u> $-\lambda < 0$) <u>if the eigenspace</u> $\mathfrak{E}_\lambda$ <u>has a finite dimension</u>. Finally we say that <u>the spectrum of the operator</u> K is discrete away from zero or simply essentially discrete if the eigenspace of each closed interval that does not contain zero has a finite dimension, i.e. if it is discrete above every positive and below every negative value of κ.

Since the operator K transforms every vector $\Phi = \eta_{\mathcal{J}}(K)\Phi$ of $\mathfrak{E}_{\mathcal{J}}$ into the vector $K\Phi = K\eta_{\mathcal{J}}(K)\Phi = \eta_{\mathcal{J}}(K)K\Phi$, which also lies in $\mathfrak{E}_{\mathcal{J}}$, it may be regarded as a Hermitean operator acting in the finite-dimensional space $\mathfrak{E}_{\mathcal{J}}$. This space, unless it is empty, is therefore spanned by a finite number of mutually orthogonal normed eigenvectors Ω. We apply this remark to the intervals $0 < \lambda < \kappa \leq ||K||$ and the corresponding eigenspaces $\mathfrak{E}_\lambda$ and conclude that every vector $\Phi^{(\lambda)}$ of this space can be written as a linear combination

(24.1) $$\Phi^{(\lambda)} = \sum_{\sigma=1}^{n_\lambda} \xi_\sigma \Omega^{(\sigma)}$$

of eigenvectors $\Omega^{(1)}, \ldots$ of the operator K such that

(24.1)' $$K\Phi^{(\lambda)} = \sum_{\sigma=1}^{n_\lambda} \kappa_\sigma \xi_\sigma \Omega^{(\sigma)} .$$

Here σ runs from 1 to n_λ, the dimension of the space $\mathfrak{E}_\lambda$, and $\kappa_1, \ldots, \kappa_n$ are the eigenvalues of K lying above λ.

We further conclude that the unit form is given by

$$(\Phi^{(\lambda)}, \Phi^{(\lambda)}) = \sum_{\sigma=1}^{n_\lambda} |\xi_\sigma|^2$$

and the associated quadratic form by

$$(\Phi^{(\lambda)}, K\Phi^{(\lambda)}) = \sum_{\sigma=1}^{n_\lambda} \kappa_\sigma |\xi_\sigma|^2 .$$

Every vector Φ in $\mathfrak{H}$ can evidently be written as the sum

(24.2) $$\Phi = \Phi^{(\lambda)} + \Phi_\lambda^{\perp}$$

of its projections $\Phi^{(\lambda)}$ and $\Phi_\lambda^{\perp}$ into the eigenspace $\mathfrak{E}_\lambda$ and the complementary space $\mathfrak{E}_\lambda^{\perp}$, so that the relations

$$(24.3) \qquad (\Phi,\Phi) = \sum_{\sigma=1}^{n_\lambda} |\xi_\sigma|^2 + (\Phi_\lambda^\perp, \Phi_\lambda^\perp) ,$$

$$(24.3)' \qquad (\Phi,K\Phi) = \sum_{\sigma=1}^{n_\lambda} \kappa_\sigma |\xi_\sigma|^2 + (\Phi_\lambda^\perp, K\Phi_\lambda^\perp)$$

hold. Note that the coefficients ξ_σ are given by the formula

$$(24.4) \qquad \xi_\sigma = (\Omega^{(\sigma)}, \Phi)$$

and are hence independent of the choice of the number λ.

Finally we state that for the vectors $\Phi_\lambda^\perp$ orthogonal to $\mathfrak{E}_\lambda$ the relation

$$(24.5) \qquad (\Phi_\lambda^\perp, K\Phi_\lambda^\perp) \leq \lambda(\Phi_\lambda^\perp, \Phi_\lambda^\perp)$$

holds.

To show this we may introduce the unit step function $\eta_\lambda^\perp(\kappa) = 1 - \eta_\lambda(\kappa)$ of the interval $\kappa \leq \lambda$ so that $\Phi_\lambda^\perp = \eta_\lambda^\perp(K)\Phi_\lambda^\perp$. Now, since evidently $(\lambda-\kappa)\eta_\lambda^\perp(\kappa) \geq 0$, we have

$$(\Phi_\lambda^\perp, (\lambda-K)\Phi_\lambda^\perp) = (\Phi_\lambda^\perp, (\lambda-K)\eta_\lambda^\perp(K)\Phi_\lambda^\perp) \geq 0$$

by rule III of the functional calculus, whence relation (24.5) follows.

Similar statements, of course, hold for the eigenspace $\mathfrak{E}_{-\lambda}$ and the spectrum of K below $-\lambda$.

Suppose now that the spectrum of K is discrete away from zero. Then we introduce the eigenspaces $\mathfrak{E}_+$ and $\mathfrak{E}_-$ of the intervals $0 < \kappa < \infty$ and $-\infty < \kappa < 0$ and the eigenspace $\mathfrak{E}_0$ of the value $\kappa = 0$. By virtue of the corollary to Lemma 2 of Section 21 in Chapter IV the projectors $\eta_\lambda^\pm(K)$ of the eigenspace $\mathfrak{E}_\lambda^\pm$ tend (strongly) to the projectors $\eta^\pm(K)$ of the spaces $\mathfrak{E}^\pm$. In other words, the projections $\Phi^{(\pm\lambda)}$ of a vector Φ into $\mathfrak{E}_\lambda^\pm$ tend to the projection $\Phi^{(\pm)}$ of Φ into $\mathfrak{E}^\pm$. Since the coefficients ξ of the projections $\Phi^{(\pm\lambda)}$ are independent of λ we obtain the following <u>statement concerning the spectral representation of an operator with an essentially</u>

discrete spectrum.

The vector Φ may be written as an infinite series

$$\Phi = \sum_{\sigma=1}^{n_+} \xi_\sigma \Omega^{(\sigma)} + \sum_{\sigma=1}^{n_-} \xi_{-\sigma}\Omega^{(-\sigma)} + \Phi_0 ,$$

where Φ_0 is the projection of Φ into $\mathfrak{E}_0$ and $\Omega^{(\sigma)}$, $\Omega^{(-\sigma)}$ are sequences of orthonormal eigenvectors of K with positive and negative eigenvalues respectively; the (finite or infinite) numbers $n^\pm$ are the dimensions of the spaces $\mathfrak{E}^\pm$. At the same time the expansion

$$K\Phi = \sum_{\sigma=1}^{n_+} \kappa_\sigma \xi_\sigma \Omega^{(\sigma)} + \sum_{\sigma=1}^{n_-} \kappa_{-\sigma}\xi_{-\sigma}\Omega^{(-\sigma)}$$

holds since $K\Phi_0 = 0$. For the unit form and the form of the operator we have

$$(\Phi,\Phi) = \sum_{\sigma=1}^{n_+} |\xi_\sigma|^2 + \sum_{\sigma=1}^{n_-} |\xi_{-\sigma}|^2 + ||\Phi_0||^2 ,$$

$$(\Phi,K\Phi) = \sum_{\sigma=1}^{n_+} \kappa_\sigma|\xi_\sigma|^2 + \sum_{\sigma=1}^{n_-} \kappa_{-\sigma}|\xi_{-\sigma}|^2 .$$

It is not implied here that there actually are positive or negative eigenvalues, i.e. that $n_+ > 0$ and $n_- > 0$. If there are none, i.e. if $n_+ = n_- = 0$, we have, of course, $K = 0$; if there are such eigenvalues $n^\pm > 0$ they can be arranged in decreasing and increasing order respectively,

$$\kappa_1 \geq \kappa_2 \geq \dots > 0; \quad \kappa_{-1} \leq \kappa_{-2} - \dots < 0 ,$$

since their number above every positive λ and below every negative $-\lambda$ is finite.

It is also not implied that there are infinitely many positive or negative eigenvalues $n_\pm = \infty$; if $n_+ = \infty$ or $n_- = \infty$ these eigenvalues tend to zero:

$$\kappa_1 \geq \kappa_2 \geq \dots \to 0 ; \quad \kappa_{-1} \leq \kappa_{-2} \leq \dots \to 0 ,$$

as follows from the fact that away from zero there is only a finite number of them.

These statements and formulas give the spectral representation of operators with an essentially discrete spectrum.

25. Completely Continuous Operators

The question naturally arises whether or not one can tell beforehand - from the nature of the operator K - that this operator has an essentially discrete spectrum. In fact it is possible to do so, as was discovered by Hilbert in 1906. Hilbert found that operators with an essentially discrete spectrum can be characterized by a simple property, which he called "complete continuity", of its form. Very frequently, it is easy to test whether or not a concretely given operator has this property.

Instead of describing Hilbert's property of complete continuity, we shall at first describe a different - but equivalent - property which, in general, is still more easily verified in concrete cases and from which the essential discreteness of the spectrum can frequently be inferred immediately. We shall call this the property of "almost finite-dimensionality". At first, however, we shall describe a less restricted property.

We shall say that the form of an operator K has dimension g above a number $\lambda > 0$ if there exist vectors $z^{(1)},\dots,z^{(g)}$ in $\mathfrak{H}$ such that the inequality

$$(25.1)_+ \qquad (\Phi,K\Phi) \leq \sum_{\gamma=1}^{g} |z^{(\gamma)},\phi|^2 + \lambda(\Phi,\Phi)$$

holds for all vectors Φ in $\mathfrak{H}$; similarly, the form will be said to have dimension g below $-\lambda < 0$ if there are vectors $z^{(-1)},\dots,$ $z^{(-g)}$ such that

$$(25.1)_- \qquad (\Phi,K\Phi) \geq - \sum_{\gamma=1}^{g} |z^{(-\gamma)},\phi|^2 - \lambda(\Phi,\Phi) .$$

Finally, we say the form of K is "almost finite dimensional" if for every $\varepsilon > 0$ there is a finite number $g = g(\varepsilon)$ of vectors $Z^{(1)}$, $\ldots, Z^{(g)}$ (also depending on ε) such that the inequality

$$(25.1)_0 \qquad |\Phi, K\Phi| \leq \sum_{\gamma=1}^{g} |Z^{(\gamma)}, \Phi|^2 + \varepsilon(\Phi, \Phi)$$

holds for every vector Φ in $\mathfrak{H}$.

Of course, the latter property could also have been described by saying that the form is almost finite-dimensional if it is finite-dimensional above every positive and below every negative number.

As seen from the two theorems that we shall prove, discreteness above λ and having finite dimension above λ are equivalent.

Theorem 1. If the spectrum of an (Hermitean bounded) operator K is discrete above $\lambda > 0$ (below $-\lambda < 0$) it is finite-dimensional above $\lambda > 0$ (below $-\lambda < 0$).

To prove this statement we write $\Phi = \Phi^{(\lambda)} + \Phi_\lambda^{\perp}$, as in Section 24, and set $Z^{(\gamma)} = \kappa_\gamma^{1/2} \Omega^{(\gamma)}$. In view of (24.5), formula (24.3)' with $g_\lambda = n_\lambda$ then assumes the form $(25.1)_+$. Similarly, one establishes $(25.1)_-$.

Theorem 2, the converse of Theorem 1, will be proved together with a corollary.

Theorem 2. If the form of an (Hermitean bounded) operator has dimension g above $\lambda > 0$ (below $-\lambda < 0$) its spectrum has dimension g above $\lambda > 0$ (below $-\lambda < 0$).

Proof. Suppose the first statement of the corollary were not true; $\dim \mathfrak{E}_\lambda > g$. Then there would exist (at least) $g+1$ linearly independent vectors $\Phi^{(1)}, \ldots, \Phi^{(g+1)}$ in $\mathfrak{E}_\lambda$. Consequently, there would exist in $\mathfrak{E}_\lambda$ a vector $\Phi \neq 0$ perpendicular to the g vectors $Z^{(1)}, \ldots, Z^{(g)}$.

For, the condition that the linear combination $\Phi = c_1\Phi^{(1)} + \ldots + c_{g+1}\Phi^{(g+1)}$ is orthogonal to $Z^{(1)}, \ldots, Z^{(g)}$ represents

g linear conditions for $g+1$ unknowns; hence there is a set of coefficients $(c_1,\ldots,c_{g+1})$, not all zero, satisfying these equations. The resulting vector Φ is not zero because of the linear independence of $\Phi^{(1)},\ldots,\Phi^{(g+1)}$. Inserting this particular vector Φ in $(25.1)_+$ we find

$$(\Phi,K\Phi) \leq \lambda(\Phi,\Phi).$$

On the other hand, since the vector $\Phi \neq 0$ is in $\mathfrak{E}_\lambda$ the inequality

$$(\Phi,K\Phi) > \lambda(\Phi,\Phi)$$

holds by virtue of Rule III of Section 21, Chapter IV. Since $(\Phi,\Phi) \neq 0$, this is a contradiction; i.e., $\dim \mathfrak{E}_\lambda \leq g_\lambda$. Similarly, one proves $\dim \mathfrak{E}_{-\lambda} \leq g_{-\lambda}$.

The main statement to be made in this section is

Theorem 3. If the form of an (Hermitean bounded) operator is almost finite-dimensional its spectrum is essentially discrete. It is an immediate consequence of Theorem 2.

As we shall see, it is in many cases easily verified that the form of an operator is almost finite-dimensional; therefore the essential discreteness of the spectrum of an operator is also easily established in many cases. Moreover, an estimate of the manner in which the eigenvalues of such an operator approach zero is given at the same time. For the corollary to Theorem 2 shows that the nth positive eigenvalue is less than or equal to λ if n is chosen greater than g_λ. In fact, one of the methods of estimating the behavior of the sequence of eigenvalues is based on just this situation.

In order to describe Hilbert's property of complete continuity, which we shall do now, we must introduce the notion of "weak convergence" of a sequence of vectors, which differs from the ordinary or "strong" convergence introduced in Section 14 of Chapter III. We say that the sequence of vectors $\Phi^{(\nu)}$ tends to zero "weakly",

$$\Phi^{(\nu)} \overset{w}{\to} 0 \, ,$$

if for every vector Ψ in a dense subspace $\mathfrak{H}'$ of $\mathfrak{H}$ the relation

$$(\Psi,\Phi^{\nu}) \to 0 \quad \text{as } \nu \to \infty \tag{25.2}$$

holds while at the same time a number C exists such that

$$||\Phi^{\nu}|| \leq C \text{ for all } \nu. \tag{25.3}$$

Clearly, it follows from these two properties that (1) holds for every* vector in $\mathfrak{H}$. For, if Ψ is any vector in $\mathfrak{H}$, there is for every $\varepsilon >$ 0 a vector Ψ' in $\mathfrak{H}'$ such that $||\Psi'-\Psi|| \leq \varepsilon/2C$ and a ν_{ε} such that $|\Psi',\Phi^{\nu}| \leq \varepsilon/2$ for $\nu \geq \nu_{\varepsilon}$, whence

$$|\Psi,\Phi^{\nu}| \leq \varepsilon \quad \text{for} \quad \nu \geq \nu_{\varepsilon} \, .$$

It follows from this remark, for example, that every sequence $\Phi^{(\nu)} = \{\xi_1^{\nu},\xi_2^{\nu},\ldots\}$ of vectors in the special Hilbert space converges weakly to zero if every component ξ^{ν} converges to zero as $\nu \to \infty$ provided $||\Phi^{\nu}|| \leq C$. For, we need only take the space of vectors with a finite number of components as subspace $\mathfrak{H}'$.

Now we may give the definition of complete continuity. The form of an Hermitean bounded operator K is said to be completely continuous if it converges to zero whenever the sequence Φ^{ν} tends to zero weakly, i.e. if

$$\Phi^{\nu} \overset{w}{\to} 0$$

implies

$$(\Phi^{\nu},K\Phi^{\nu}) \to 0 \, .$$

---

*In fact, this stronger version of property (1) implies property (2), by virtue of the principle of uniform boundedness (due to Hellinger and Toeplitz); but we also do not need this fact.

We then formulate

Theorem 4. If the form of the (Hermitean bounded) operator K is almost finite-dimensional, it is completely continuous.

Proof. Let $\Phi^{\nu} \to 0$. To any $\varepsilon > 0$ take vectors $Z^{(1)},\ldots,Z^{(g)}$ such that inequality $(25.1)_0$ holds. Then a $\nu = \nu_{\varepsilon}$ can be found such that $|Z^{(\gamma)},\Phi^{\nu}|^2 \leq \varepsilon$ for $\nu \geq \nu_{\varepsilon}$. By $(25.1)_0$, therefore,

$$|\Phi^{\nu},K\Phi^{\nu}| \leq (g+C)\varepsilon \ ;$$

hence the statement is proved.

The converse of Theorem 4 also holds, as could be shown by deriving the statement of Theorem 3 directly from the complete continuity of the form of K. This converse could also be proved directly. We find it preferable to take the property of being almost finite-dimensional as the basic one. A related property was already used by Hellinger and Toeplitz, although they did not give this property a special name.

The notion of complete continuity is, however, very useful if one wants to exhibit counterexamples.

Remark. If the form of the (Hermitean bounded) operator K is not completely continuous its spectrum is not essentially discrete.

This statement follows immediately from Theorem 4 by combining it with the converse of Theorem 3, which is implied by Theorem 1.

We shall use this remark in Section 25 to show that certain operators do not have an essentially discrete spectrum.

We should mention that there are several different formulations of complete continuity which are all equivalent when they refer to a Hilbert space. F. Riesz has given the following two striking such formulations.

1. The operator K is completly continuous if it transforms any

weakly convergent sequence into a strongly convergent one.

2. The operator K is completely continuous if it transforms any bounded subset of the space into a set whose closure is compact. (It is now customary to call an operator "compact" if it has the last property.)

Note that in the last formulation no reference to an inner product is made (this reference could also be eliminated from the first one by replacing the inner product by a bounded linear functional in the definition of weak convergence). It is thus understandable that these formulations are important in the theory of operators in spaces more general than the Hilbert space. However, in the work presented in these notes we shall have no occasion to employ these formulations.

26. Completely Continuous Integral Operators

Before showing that integral operators of a wide class are completely continuous, we make a general remark which will be helpful in this connection.

We say a Hermitean operator K is "of finite rank" if there is a number of vectors $z^1,\dots,z^g$ and numbers $k_{\lambda\lambda'} = k_{\lambda'\lambda}$, for $\lambda,\lambda' = 1,\dots,g$, such that

$$K\Phi = \sum_{\lambda,\lambda'} k_{\lambda\lambda'}(z^{\lambda'},\Phi)\, z^{\lambda} . \tag{26.1}$$

Evidently, the range of such an operator is finite-dimensional. Next we say a Hermitean operator K is "almost of finite rank" if for every $\varepsilon > 0$ there is an operator K of finite rank such that

$$||K^{\varepsilon}-K|| \leq \varepsilon . \tag{26.2}$$

Here we have employed the "operator norm" of an operator. If such an estimate holds for any other norm, it certainly holds for the minimal norm. Using these definitions we make the

Remark. If a Hermitean operator K is almost of finite rank its form is almost finite-dimensional.

To prove this statement we let K^{ε} be of the form (26.1); then we have

$$|\Phi,K^{\varepsilon}\Phi| \leq \sum_{\lambda,\lambda'} |k_{\lambda\lambda'}| \, |z^{\lambda'},\Phi| \, |z^{\lambda},\Phi|$$

$$\leq \max_{\lambda} \sum_{\lambda'} |k_{\lambda\,\lambda'}| \sum_{\lambda} |z_{\lambda},\Phi|^2$$

Inserting this into $|\Phi,K\Phi| \leq |\Phi,K^{\varepsilon}\Phi| + \varepsilon|\Phi,\Phi|$ we obtain the statement $(25.1)_0$ (except for an irrelevant factor).

We recall from Section 20 in Chapter IV that an integral operator K with the kernel $k(s,s')$ assigns the function

$$K\phi(s) = \int k(s,s')\phi(s')dr(s')$$

to the function $\phi(s)$. Here $r(s)$ is a non-decreasing (real) measure function.

Let us first assume the kernel $k(s,s')$ to be piecewise continuous with reference to a partition $\mathcal{P} \times \mathcal{P}$ of the (s,s')-plane which is the product of a partition $\mathcal{P}$ of the s-axis with itself. I.e., let $\mathcal{J}_\sigma$ stand for the open and closed cells of the partition $\mathcal{P}$; then the partition $\mathcal{P} \times \mathcal{P}$ of the (s,s')-plane is given by the product $\mathcal{J}_\sigma \times \mathcal{J}_{\sigma'}$. The kernel $k(s,s')$ is piecewise constant if in each such product cell it agrees with a function which is continuous in the closure of this cell.

For any piecewise continuous function $\phi(s)$ with bounded support the function $K\phi(s)$ is defined and piecewise continuous. In Section 20, Chapter IV, these functions $\phi(s)$ were extended to a Hilbert space with the unit form

$$||\phi||^2 = \int |\phi(s)|^2 dr(s) .$$

Suppose now the kernel $k(s,s')$ has a finite bound, such as the Hilbert-Schmidt bound or the Holmgren bound. Then the operator K is bounded and can be extended to the whole Hilbert space of functions $\phi(s)$. Certainly the kernel k, and hence the operator K, is bounded if this kernel has bounded support.

We now state

<u>Theorem</u> 26.1. An integral operator K having a piecewise continuous kernel with bounded support is almost finite-dimensional and hence completely continuous.

Let us first assume that the kernel $k(s,s')$ is piecewise constant and of bounded support. It can then be written in the form

$$k(s,s') = \sum_{\lambda\lambda'} k_{\lambda\lambda'} \eta_{\lambda'}(s')\eta_{\lambda}(s) ,$$

which shows that the operator K is of finite rank. Note that here the number of terms is finite since $k(s,s')$ was assumed of bounded support.

Next we let the kernel $k(s,s')$ be piecewise continuous and of bounded support. Clearly, such a kernel can be approximated by kernels which are piecewise constant with reference to appropriate subdivisions of the partition $\mathscr{P} \times \mathscr{P}$. For every $\varepsilon > 0$ we therefore can find a piecewise constant kernel $k^{\varepsilon}(s,s')$ chosen such that $\max_{s,s'}|k^{\varepsilon}(s,s') - k(s,s')|$ is so small that $||k^{\varepsilon}-k||_1 \leq \varepsilon$ and hence $||K^{\varepsilon}-K|| \leq \varepsilon$; see Section 20. The remark made at the beginning of Section 26 then gives the statement of the theorem.

This theorem immediately leads to the following

<u>Corollary</u>. An integral operator which can be uniformly approximated by an integral operator whose kernel is piecewise continuous with bounded support is almost finite-dimensional and hence completely continuous.

For, since K can be approximated by an operator K' which is almost of finite rank such that $||K'-K||$ is arbitrarily small, it is itself almost of finite rank and hence the "remark" is again applicable.

To be sure, the class of integral operators thus covered is very large. It does not only comprise integral operators wich continuous kernels defined in a finite region; it comprises also certain integral operators in infinite regions whose kernels may have certain singularities; the criterion is whether or not these kernels can be approximated in norm by non-singular kernels in finite regions. Also infinite matrices are covered, since the measure function $r(s)$ is permitted to be constant except for a sequence of jumps.

For all the operators thus covered a spectral representation is furnished by the results of Section 24. For all such operators there exist (infinite, finite, or absent) sequences of orthonormal eigenvectors $\Omega^{(\pm 1)}, \Omega^{(\pm 2)}, \ldots$ with eigenvalues $\kappa_{\pm 1}, \kappa_{\pm 2}, \ldots \lessgtr 0$ which tend to zero if there are infinitely many of them. In addition, $\kappa = 0$ may be an eigenvalue of finite or infinite multiplicity.

Of course, not all integral operators are covered by the class described. There are integral operators K which can be approximated by operators K of finite rank in the strong sense and still are not completely continuous. It then follows that these operators cannot be approximated uniformly by operators of finite rank.

A typical example is any integral operator H whose kernel $h(s,s')$ is generated by a single function $h(s)$ in the form

$$h(s,s') = h(s-s')$$

while the measure function is simply $r(s) = s$. We assume $h(s)$ to be piecewise continuous and such that

$$\int |h(s)| ds < \infty ;$$

then, evidently, the Holmgren norm is finite and hence the operator K is bounded.

Let $\eta_a(s)$ be the unit step function of the interval $|s| < a$. Then, we maintain, the operator H^a with the kernel

$$h^a(s,s') = \eta_a(s)h(s-s')\eta_a(s')$$

approximates the operator K strongly. To prove this, it is sufficient to show (1) that the operator H^a is bounded independently of a, and (2) that $H^a\phi$ tends to $H\phi$ for a dense subspace of functions ϕ. The boundedness of H^a is clear since $||h^a||_1 \leq ||h||_1$.

The convergence $H^a\phi \to H\phi$ is evident for functions ϕ of bounded support since $H^a\phi(s) = \eta_a(s)H\phi(s)$ for these functions provided a is chosen sufficiently large.

Now, we can easily show that <u>our operator</u> H <u>is not completely continuous</u>. At this place it is advantageous to use Hilbert's definition of this notion. To show that Hilbert's property is violated we need only exhibit a sequence of functions ϕ^ν in $\mathfrak{H}$ which converges weakly to zero while $(\phi^\nu, H\phi^\nu)$ does not converge to zero.

To this end we take one particular function $\phi(s)$ with support in $|s| \leq a$ and for which

$$\iint \phi(s)h(s-s')\phi(s')ds'ds \neq 0 .$$

Then we set

$$\phi^\nu(s) = \phi(s-\nu) .$$

For these functions, clearly

$$||\phi^\nu|| = ||\phi||$$

and

$$\int \psi^b(s)\phi^\nu(s)ds = \int \psi^b(s)\phi(s-\nu)ds = 0$$

if $\nu > a+b$ provided the function ψ^b in $\mathfrak{H}$ has support in $|s| \leq b$. Since the space of these functions ψ^b is dense in $\mathfrak{H}$, it follows that ϕ^ν converges weakly to zero. On the other hand, setting

$$H\phi(s) = \int h(s-s')\phi(s')ds'$$

we find

$$H\phi^\nu(s) = \int h(s-s'-\nu)\phi(s')ds'$$

and

$$(\phi^\nu, H\phi^\nu) = \int\int \phi(s-\nu)h(s-\nu-s')\phi(s')ds'ds$$

$$= \int\int \phi(s)h(s-s')\phi(s')ds'ds \neq 0$$

independently of ν. Therefore $(\phi^\nu, H\phi^\nu)$ does not converge to zero. Thus it is shown that the operator H is not completely continuous; its spectrum is not essentially discrete.

27. Maximum - Minimum Properties of Eigenvalues

If the spectrum of an Hermitean operator K above a value $\lambda > 0$ is discrete and not empty, so that it possesses a largest eigenvalue, this eigenvalue can be characterized as the maximum of the quadratic form $(\Phi, K\Phi)$ taken for all vectors Φ with $||\Phi|| = 1$. This fact is evident from the formulas

$$(\Phi, K\Phi) = \sum_{\sigma=1}^{n} K_\sigma |\eta_\sigma|^2 + (\Phi_\lambda, K\Phi_\lambda) ,$$

$$(\Phi,\Phi) = \sum_{\sigma=1}^{n} |\eta_\sigma|^2 + (\Phi_\lambda^\perp,\Phi_\lambda^\perp)$$

in conjunction with $(\Phi_\lambda^\perp, K\Phi_\lambda^\perp) \leq \lambda(\Phi_\lambda^\perp,\Phi_\lambda^\perp)$ given in Section 24. For, with κ_1 taken as the largest eigenvalue we deduce from them the relation

$$(\Phi,K\Phi) - \kappa_1(\Phi,\Phi) \leq - \sum_{\sigma=2}^{n} (\kappa_1-\kappa_\sigma)|\eta_\sigma|^2 - (\kappa_1-\lambda)\,||\Phi_\lambda^\perp||^2 \leq 0,$$

the inequality being assumed for $\Phi = \Omega^{(1)}$. Hence the statement follows.

<u>Theorem</u> 27.1. The mth <u>eigenvalue</u> κ_m in the sequence $\kappa_1 \geq \kappa_2 \geq \cdots \geq \kappa_n$ can be characterized as <u>the maximum of the quadratic form</u> $(\Phi,K\Phi)$ for all vectors Φ with norm 1 which are orthogonal to $\Omega^{(1)},\ldots,\Omega^{(m-1)}$.

For, with such a vector Φ the relation

$$(\Phi,K\Phi) - \kappa_m(\Phi,\Phi) \leq - \sum_{\sigma=m+1}^{n} (\kappa_m-\kappa_\sigma)|\eta_\sigma|^2 - (\kappa_m-\lambda)\,||\Phi_\lambda^\perp||^2 \leq 0$$

holds, the equality being assumed for $\Phi = \Omega^{(m)}$.

It is an important fact that the mth eigenvalue can also be characterized as a minimum without reference to the m-1 first eigenvectors. This fact is expressed by the

<u>Theorem</u> 27.2. Suppose the eigenspace $\mathfrak{E}_\lambda$, $\lambda > 0$ on the (Hermitean bounded) operator K has the dimension $n \geq m$. Then <u>the</u> mth <u>eigenvalue of</u> K <u>is the minimum with respect to the choice of</u> m-1 <u>vectors</u> $X^{(1)},\ldots,X^{(m-1)}$ <u>of the "maximum" of the quadratic form</u> $(\Phi,K\Phi)$ <u>taken for vectors</u> Φ <u>with the norm</u> 1, <u>orthogonal to</u> $X^{(1)},\ldots,X^{(m-1)}$. (By "maximum" here we mean the "least upper bound" since we do not intend to prove that an actual maximum is assumed.)

Evidently, one can choose a vector $\Phi = \eta_1\Omega^{(1)} + \ldots + \eta_m\Omega^{(m)} \neq 0$ spanned by the m eigenvectors $\Omega^{(1)},\ldots,\Omega^{(m)}$ which is orthogonal to any chosen m-1 vectors $X^{(1)},\ldots,X^{(m-1)}$, since the determination of such a vector involves the solution of $m-1$ homogeneous equations for m unknowns.

The value of the ratio $(\Phi,K\Phi)/(\Phi,\Phi)$ for this vector is evidently $\geq \kappa_m$, the smallest of the eigenvalues $\kappa_1,\ldots,\kappa_m$. The same is hence true of the "maximum" of this ratio for vectors $\Phi \perp X^{(1)},\ldots, X^{(m-1)}$. Since this "maximum" equals κ_m for $X^{(1)} = \Omega^{(1)},\ldots, X^{(m-1)} = \Omega^{(m-1)}$, as observed above, the value κ_m is indeed seen to be the minimum of this "maximum".

The fact stated in Theorem 27.1 enables one to study the effect which a change of the operator K has on its eigenvalues. For differential operators the corresponding fact was derived and widely employed by Courant.

Another, complementary way of characterizing the mth eigenvalue should be mentioned.

Theorem 27.3. Suppose the eigenspace $\mathfrak{E}_\lambda$, $\lambda > 0$, of the (Hermitean bounded) operator K has the dimension $n \geq m$. Then *the mth eigenvalue of K is the maximum with respect to the choice of m linearly independent vectors $\Xi^{(1)},\ldots,\Xi^{(m)}$ of the "minimum" of the quadratic form $(\Phi,K\Phi)$ taken for vectors Φ with the norm* 1, which are linear combinations of $\Xi^{(1)},\ldots,\Xi^{(m)}$.

To establish this fact we observe that there is at least one such combination $\Phi \neq 0$ which is orthogonal to $\Omega^{(1)},\ldots,\Omega^{(m-1)}$. For this vector Φ we have $(\Phi,K\Phi)/(\Phi,\Phi) \leq \kappa_m$. The same is, therefore, true of the "minimum" of the ratio. For $\Xi^{(1)} = \Omega^{(1)},\ldots, \Xi^{(m)} = \Omega^{(m)}$, now, this minimum evidently equals κ_m. Consequently, κ_m is the maximum of the "minimum".

The maximum (minimum) property of positive (negative) eigenvalues can be used to derive the spectral resolution of almost finite-dimensional operators without relying on the general spectral theory of bounded operators developed in Chapter IV. We first prove the existence of a largest eigenvalue for an operator which is finite-dimensional above a number $\lambda > 0$ under a simple condition.

Theorem 27.4. Suppose the form of the selfadjoint bounded operator K is (I) finite-dimensional above a number $\lambda > 0$. Furthermore, assume (II) that a vector $\Phi_0 \neq 0$ exists such that

$$(\Phi_0, K\Phi_0) > \lambda(\Phi_0, \Phi_0) \; . \tag{27.1}$$

Then the operator K possesses a largest eigenvalue $\kappa_1 > \lambda$. It is given as the maximum of the ratio $(\Phi, K\Phi)/(\Phi, \Phi)$ for $\Phi \neq 0$.

To prove it, let κ_1 be the least upper bound of this ratio and let Φ^ν be a sequence with $||\Phi^\nu|| = 1$ for which

$$(\Phi^\nu, K\Phi^\nu) \to \kappa_1 \quad \text{as} \quad \nu \to \infty. \tag{27.2}$$

From it we may select a subsequence - also denoted by Φ^ν - such that each of the g inner products (Z^γ, Φ^ν) tends to a limit, where $Z^{(1)}, \ldots, Z^{(g)}$ are the vectors figuring in the inequality $(25.1)_+$ which expresses the hypothesis (I) that $(\Phi, K\Phi)$ be finite-dimensional above λ. Introducing the difference

$$\Phi^{\nu\mu} = \Phi^\nu - \Phi^\mu$$

we may express this requirement by the relation

$$(Z^\gamma, \Phi^{\nu\mu}) \to 0 \quad \text{as } \nu, \mu \to \infty \quad \text{for } \gamma = 1, \ldots, g \; .$$

We maintain that the subsequence Φ^ν so chosen is a Cauchy sequence.

To prove this we first note that the quadratic form $(\Phi, (\kappa_1 - K)\Phi) = \kappa_1(\Phi, \Phi) - (\Phi, K\Phi)$ is non-negative, by virtue of the definition of κ_1. Hence we may employ the same identity which we have employed in proving the projection theorem:

$$((\Phi^\nu-\Phi^\mu),(\kappa_1-K)(\Phi^\nu-\Phi^\mu)) + ((\Phi^\nu+\Phi^\mu),(\kappa_1-K)(\Phi^\nu+\Phi^\mu))$$

$$= 2(\Phi^\nu,(\kappa_1-K)\Phi^\nu) + 2(\Phi^\mu,(\kappa_1-K)\Phi^\mu) .$$

With $\Phi^\nu-\Phi^\mu = \Phi^{\nu\mu}$ we derive from it

$$(\Phi^{\nu\mu},(\kappa_1-K)\Phi^{\nu\mu}) \leq 2(\Phi^\nu,(\kappa_1-K)\Phi^\nu) + 2(\Phi^\mu,(\kappa_1-K)\Phi^\mu) .$$

To this inequality we add the inequality $(25.1)_+$ for $\Phi^{\nu\mu}$, which we may write in the form

$$(\Phi^{\nu\mu},(K-\lambda)\Phi^{\nu\mu}) \leq \sum_{\gamma=1}^{g} |Z^{(\gamma)},\Phi^{\nu\mu}|^2$$

in obvious notation. Thus we obtain

$$(\kappa_1-\lambda)(\Phi^{\nu\mu},\Phi^{\nu\mu}) \leq 2(\Phi^\nu,(\kappa_1-K)\Phi^\nu) + 2(\Phi^\mu,(\kappa_1-K)\Phi^\mu) + \sum_{\gamma=1}^{g} |Z^{(\gamma)},\Phi^{\nu\mu}|^2 .$$

We maintain that all terms on the right-hand side tend to zero as $\nu,\mu \to \infty$. For the first two terms this follows from relation (27.1), for the last term from relation (27.2).

From this fact we may conclude

$$||\Phi^{\nu\mu}|| \to 0 ,$$

since $\kappa_1 > \lambda$ because of hypothesis II and the definition of κ_1. Thus we have proved that Φ^ν is a Cauchy sequence.

By virtue of the completeness of the Hilbert space there is a vector $\Omega^{(1)}$ such that

$$\Phi^\nu \to \Omega^{(1)} ;$$

furthermore,

$$||\Omega^{(1)}|| = 1$$

and

$$(\Omega^{(1)},(\kappa_1-K)\Omega^{(1)}) = 0 .$$

From the last relation we may conclude

$$(\kappa_1 - K)\,\Omega^{(1)} = 0$$

by Theorem 19.1, since the form of $\kappa_1 - K$ is non-negative. Thus it follows that κ_1 is an eigenvalue and $\Omega^{(1)}$ an eigenvector of K.

To establish the existence of the second eigenvalue we need only apply Theorem 27.3 to the space $\mathfrak{H}_1^{\perp}$ orthogonal to $\Omega^{(1)}$, observing that $K\Phi$ is in this space if Φ is, as follows from

$$(\Omega^{(1)}, K\Phi) = (K\Omega^{(1)}, \Phi) = \kappa_1(\Omega^{(1)}, \Phi) \ .$$

Of course, this theorem is only applicable if hypothesis II holds also in $\mathfrak{H}_1^{\perp}$.

Continuing in this fashion we may establish the existence of a non-increasing sequence of eigenvalues, as long as hypothesis II remains valid. This hypothesis ceases to be valid before the $(g+1)$st step. For, if there were $g+1$ eigenvalues $> \lambda$ the ratio $(\Phi, K\Phi)/(\Phi, \Phi)$ would be $> \lambda$ for any linear combination $\Phi \neq 0$ of the first $g+1$ eigenvectors, while such a combination exists which is orthogonal to the z^{γ} and hence contradicts this statement by virtue of inequality $(25.1)_-$.

If the form of the operator K is almost finite-dimensional one obtains in the manner described a sequence of non-increasing positive and non-decreasing negative eigenvalues which tend to zero if there are infinitely many of them. The spectral representation described in Section 23 can then be established. We shall not describe the rather obvious steps needed to complete this argument.

CHAPTER VI

NON-BOUNDED OPERATORS

28. Closure and Adjointness

Operators which are not bounded will not be defined in the whole Hilbert space $\mathfrak{H}$, but only in a subspace of $\mathfrak{H}$, called the "domain" of the operator and denoted by $\mathfrak{H}_A$. Operators defined only in a subspace of $\mathfrak{H}$ occur quite frequently. Of course, integral operators are naturally defined only in such subspaces of $\mathfrak{H}$. But if they are bounded - and those that we have considered are bounded - they can be extended to the whole space $\mathfrak{H}$. Differential operators also are defined only in subspaces - as was already indicated in Chapter I; but they are always strictly non-bounded, as will be shown in Chapter VII. Of course one will naturally try to extend the domain of a non-bounded operator as far as possible.

In this section we shall discuss procedures for extending an operator defined in a subspace of $\mathfrak{H}$ as far as possible - in a sense to be defined. If the operator is actually non-bounded these extended domains are actually subspaces of the Hilbert space.* In these procedures we shall use the notions of closure and of "formal" and "strict" adjointness. These notions are important aside from their role in extension procedures. In particular, we mention that "strict self-adjointness" of an operator is the property which makes its spectral representation possible, as we shall show in Section 30.

In the discussions of the present and some later sections it is advisable to extend the notion of "operator" slightly by allowing

*It should be mentioned that a Hermitean operator which is applicable on every vector in a Hilbert space is automatically bounded, provided it is closed (see below). This remarkable fact represents one of the deep results in the theory of operators. Still we shall not have occasion to use it; for, in order to exhibit specific operators that are applicable on all of Hilbert space, we shall always presuppose the existence of a bound; see Section 20.

the range of an operator A to lie in a Hilbert space $\mathfrak{G}$, which need not be the same as the original space $\mathfrak{H}$.

First we introduce the notion of "closedness" and "closure". To this end we start out with an operator A defined in a domain $\mathfrak{H}_A \subset \mathfrak{H}$. Then we form the space $\overline{\mathfrak{H}_A}$ of all vectors Φ_0 in $\mathfrak{H}$ that can be approximated (strongly) by a sequence of vectors Φ^ν in $\mathfrak{H}_A$ such that the sequence $A\Phi^\nu$ converges (strongly) to a vector Φ_1 in $\mathfrak{H}$.

If $\overline{\mathfrak{H}_A} = \mathfrak{H}_A$, i.e. if each such vector Φ_0 is in $\mathfrak{H}_A$, and if $\Phi_1 = A\Phi_0$, we say that the space $\mathfrak{H}_A$ is <u>closed</u>.

If $\mathfrak{H}_A$ is not closed we may try to extend the operator A to an operator $\bar{A}$ defined in $\overline{\mathfrak{H}_A}$ by setting

$$\bar{A}\Phi_0 = \Phi_1 .$$

That can be done if Φ_1 depends uniquely and linearly on the vector Φ_0, independently of the approximating sequence. In that case we call the operator $\bar{A}$ the <u>closure</u> of A in $\mathfrak{H}_A$. The space $\mathfrak{H}_{\bar{A}} = \overline{\mathfrak{H}_A}$ is then the domain of $\bar{A}$. Clearly, $\bar{A}$ is closed in $\mathfrak{H}_{\bar{A}}$.

Below (in Theorem 28.3) we shall give a condition under which a closure exists. Before doing so we must describe the notions of adjointness.

In accordance with the definition given in Section 9, we say that an operator $A^\dagger$ defined in a domain $\mathfrak{G}_{A^\dagger}$ is "formally adjoint" to A in $\mathfrak{H}_A$ if the relation

$$(\dagger) \qquad (A^\dagger\Psi,\Phi) = (\Psi,A\Phi)$$

holds for all Φ in $\mathfrak{H}_A$, Ψ in $\mathfrak{G}_{A^\dagger}$.

We furthermore say that the operator A^* defined in the domain $\mathfrak{G}_{A^*}$ is the "strict" adjoint of A in $\mathfrak{H}$ if it is firstly formally adjoint to A in $\mathfrak{H}_A$,

$(*)_1$ $\qquad (A^*\Psi,\Phi) = (\Psi,A\Phi)$ holds for Φ in $\mathfrak{H}_A$, Ψ in $\mathfrak{C}_{A^*}$,

and in addition has the following "second property": whenever two vectors Ψ in $\mathfrak{C}$, Ψ_1 in $\mathfrak{H}$ are such that the relation

$(*)_2$ $$(\Psi_1,\Phi) = (\Psi,A\Phi)$$

holds for all Φ in $\mathfrak{H}_A$, the vector Ψ is in $\mathfrak{C}_{A^*}$ and $A^*\Psi = \Psi_1$.

This second property may be expressed by saying that whenever a vector Ψ together with a vector Ψ_1 satisfies the relation that would hold if Ψ were in $\mathfrak{C}_{A^*}$ and Ψ_1 were equal to $A^*\Psi$, then Ψ is in $\mathfrak{C}_{A^*}$ and $\Psi_1 = A^*\Psi$. Note that the domain of A^* must be taken sufficiently large for this second property to obtain, while it might be necessary to narrow down this domain to insure that the first property, $(*)_1$, obtains.

Without giving the obvious proof, we note the fact that <u>the strict adjoint</u> A^* <u>of an operator</u> A <u>is closed in its domain</u> $\mathfrak{C}_{A^*}$. This fact illustrates the "wideness" of this domain.

If the domain of both A and its formal adjoint $A^\dagger$ are the full spaces $\mathfrak{H}$ and $\mathfrak{C}$, the operator $A^\dagger$ is strictly adjoint to A. For, relation $(*)_2$ then leads to the relation $(\Psi_1,\Phi) = (A^*\Psi,\Phi)$ for all Φ in $\mathfrak{H}$ whence $A^*\Psi = \Psi_1$. This remark shows in particular that <u>bounded formally adjoint operators defined in</u> $\mathfrak{H}$ <u>are strictly adjoint</u>. The qualification "formal" or "strict" can therefore be omitted for bounded operators.

The following theorem gives a simple condition under which the operator A has a strict adjoint.

<u>Theorem</u> 28.1. Suppose the domain $\mathfrak{H}_A$ of the operator A is dense in $\mathfrak{H}$. Then A possesses a strict adjoint.

Proof. Let Ψ be a vector in $\mathfrak{G}$ to which there is a Ψ_1 in $\mathfrak{H}$ such that relation $(*)_2$ holds for all Φ in $\mathfrak{H}_A$. Call $\mathfrak{G}_{A*}$ the space formed by all these vectors Ψ_1. Suppose $\Psi = 0$, then $(\Psi_1,\Phi) = 0$ for all Φ in $\mathfrak{H}_A$; and since this space was assumed to be dense we conclude $\Psi_1 = 0$. Consequently, the assignment of Ψ_1 to Ψ is unique and hence constitutes a linear operator, called A^*, with the domain $\mathfrak{H}_{A*}$. By its very definition this operator has the two properties characterizing it as the desired strict adjoint.

We mention that the converse also holds: if A in $\mathfrak{H}_A$ has a strict adjoint, $\mathfrak{H}_A$ is dense in $\mathfrak{H}$. We omit the easy proof.

We can use Theorem 28.1 for an extension of the operator A. To this end we apply this theorem to a formal adjoint of A.

Theorem 28.2. Suppose the operator A in $\mathfrak{H}_A$ has a formal adjoint $A^\dagger$ with a dense domain $\mathfrak{G}_{A^\dagger}$. Then there exists to $A^\dagger$ a strict adjoint $A^{\dagger *}$ in a domain $\mathfrak{H}_{A^{\dagger *}}$ which is an extension of A in $\mathfrak{H}_A$. In other words, $\mathfrak{H}_A \subset \mathfrak{H}_{A^{\dagger *}}$ and $A^{\dagger *}\Phi = A\Phi$ for Φ in $\mathfrak{H}_A$. The existence of the operator $A^{\dagger *}$, the "strict adjoint of the formal adjoint", follows from Theorem 29.1. The second property, expressing this strict adjointness, now assumes the following form. Relation

$$(*)_2^* \qquad (A^\dagger\Psi,\Phi) = (\Psi,\Phi_1) \quad \text{for all } \Psi \text{ in } \mathfrak{G}_{A^\dagger}$$

implies Φ is in $\mathfrak{H}_{A^{\dagger *}}$ and $A^{\dagger *}\Phi = \Phi_1$. Now, any Φ in $\mathfrak{H}_A$ with $\Phi_1 = A\Phi$ does satisfy $(*)_2^*$ because of relation $(\dagger)$. Hence Φ is in $\mathfrak{H}_{A^{\dagger *}}$ and $A^{\dagger *}\Phi = A\Phi$. Thus the theorem is proved.

Note that the operator $A^{\dagger *}$ is a closed extension of A (in view of the remark made above that any strict adjoint is closed), but need not be its closure. It is remarkable that the closure exists under the same circumstances under which the operator $A^{\dagger *}$ exists.

Theorem 28.3. Suppose the operator A in $\mathfrak{H}_A$ has a formal adjoint $A^\dagger$ with a dense domain $\mathfrak{C}_{A^\dagger}$; then it possesses a closure, $\bar{A}$ in $\mathfrak{H}_{\bar{A}}$.

We define $\mathfrak{H}_{\bar{A}}$ as the space of all Φ in $\mathfrak{H}$ which can be approximated by vectors Φ^ν in $\mathfrak{H}_A$ such that $A\Phi^\nu$ tends to a vector Φ_1. These vectors Φ, Φ_1 certainly satisfy condition $(*)_2^*$ since by $(\dagger)$ this relation holds for $\Phi = \Phi^\nu$ and $\Phi_1 = A\Phi^\nu$. It now follows from Theorem 28.2 that Φ admits $A^{\dagger *}$ and that $A^{\dagger *}\Phi = \Phi_1$. In other words, $\mathfrak{H}_{\bar{A}} \subset \mathfrak{H}_{A^{\dagger *}}$. Therefore, the assignment of Φ_1 to Φ in $\mathfrak{H}_{\bar{A}}$ constitutes an operator, $\bar{A}$, the restriction of $A^{\dagger *}$ to $\mathfrak{H}_{\bar{A}}$. Clearly this operator $\bar{A}$ is the desired closure.

Under the hypotheses of Theorems 28.2,3 we thus can form extensions of the operator A in $\mathfrak{H}_A$ to closed operators in two ways: the extension $\bar{A}$, also called "minimal" or "strong" extension (since strong convergence was used in its formation), and the extension $A^{\dagger *}$, also called "maximal" or "weak" extension (since the statement that a vector admits $A^{\dagger *}$ is weaker than the statement that it admits $\bar{A}$).

The relationship between the corresponding domains is indicated by

$$\mathfrak{H}_A \subset \mathfrak{H}_{\bar{A}} \subset \mathfrak{H}_{A^{\dagger *}} \subset \mathfrak{H} .$$

Naturally the question arises whether or not the strong and the weak extensions agree. That this need not be the case will be shown in connection with differential operators.

Note that the weak extension $A^{\dagger *}$ depends on the choice of the formal adjoint $A^\dagger$. If the domain of this adjoint is enlarged, the domain of $A^{\dagger *}$ is possibly reduced. The maximal enlargement of the domain of $A^\dagger$ is reached by taking the strict adjoint itself (assuming it exists). The adjoint A^{**} of A^* furnishes the strongest possible weak extension of A; it is a remarkable fact that it agrees with the strong extension of A.

Theorem 28.4. Suppose the operator A in $\mathfrak{H}_A$ has a strict adjoint A^* with a domain $\mathfrak{G}_{A^*}$ dense in $\mathfrak{G}$. Then the strict adjoint A^{**} of A^* is the closure $\bar{A}$ of A in $\mathfrak{H}_A$. That is, the strongest weak extension A^{**} of A is the strong extension $\bar{A}$ of A.

The formulation of this theorem and its proof (as all notions developed in this and many other sections) are due to von Neumann. The proof is based on the notion of "graph" of an operator.

The space of pairs $\{\Phi,\Psi\}$ of vectors in $\mathfrak{H}$ and $\mathfrak{G}$ again forms a Hilbert space $\mathfrak{H} \oplus \mathfrak{G}$ with respect to the unit form $||\Phi||^2 + ||\Psi||^2$. To every operator A in $\mathfrak{H}_A$ one may assign as its graph the subspace of $\mathfrak{H} \oplus \mathfrak{G}$ formed by the pairs $\{\Phi,A\Phi\}$ with Φ in $\mathfrak{H}_A$. If $A = \bar{A}$ is closed, then so is obviously the graph. To an operator A' acting in a subspace $\mathfrak{G}'$ of $\mathfrak{G}$ we assign as its graph the subspace of all pairs of the pairs $\{-A'\Psi,\Psi\}$ with Ψ in $\mathfrak{G}_{A'}$. The statement that $A^\dagger$ is a formal adjoint to A can then simply be described by saying that the graph of A is perpendicular to the graph of $A^\dagger$ in $\mathfrak{G}_{A'}$. Moreover, the operator A^* in $\mathfrak{H}_{A^*}$ is the strict adjoint of A in $\mathfrak{H}_A$ if its graph consists exactly of those vectors $\{-\Psi_1,\Psi\}$ which are perpendicular to the graph of A in $\mathfrak{H}_A$. By virtue of the projection theorem the orthogonal space of the orthogonal space of a subspace is exactly this subspace if it was closed to begin with. In other words, the contention $A^{**} = \bar{A}$ is proved.

An operator will be called "strictly self adjoint" in a dense domain if it is equal to its own strict adjoint: $\mathfrak{G} = \mathfrak{H}$, $\mathfrak{G}_{A^*} = \mathfrak{H}_A$, $A^* = A$. The question then will arise how one can tell whether or not a formally self-adjoint operator is strictly self-adjoint, or can be extended to a strictly self-adjoint operator. Instead of trying to answer this question directly, we shall show that strictly self-adjoint operators can be built up with the aid of a pair of strictly adjoint operators. In doing this it is convenient to make use of some theorems on non-negative forms.

29. Closed Forms

A bilinear form $(\Phi',\Phi)_1$ defined for vectors Φ,Φ' in a sub-space $\mathfrak{H}_1$ of $\mathfrak{H}$ is said to be "Hermitean" if it has the properties (7.1,2) of an inner product (linear in Φ, anti-linear in Φ'); it is said to be non-negative if it has property (7.3), $(\Phi,\Phi)_1 \geq 0$ for Φ in $\mathfrak{H}_1$; it is said to be a "sub-inner-product" if $(\Phi,\Phi)_1 \geq (\Phi,\Phi)$ holds in $\mathfrak{H}_1$. Then $[(\Phi,\Phi)_1]^{1/2}$ will be called a "sub-norm". If the space $\mathfrak{H}_1$ is complete with respect to this sub-norm the sub-norm, as well as the sub-inner-product, will be called closed.

In the present section we shall assume the sub-norm to be closed. In Section 31 we shall give various conditions under which the sub-norm can be extended to a larger domain so that is closed there. At present we prove the fundamental

Theorem 29.1. Let $||\Phi||_1$ be a closed sub-norm defined in a space $\mathfrak{H}_1$ which is dense in $\mathfrak{H}$. Then there exists a sub-space $\mathfrak{H}_2$ dense in $\mathfrak{H}_1$ with respect to $|| \ ||_1$ in which an operator F can be defined such that

$$(\Phi_1',F\Phi_2) = (\Phi_1',\Phi_2)_1 \tag{29.1}$$

for all Φ_1' in $\mathfrak{H}_1$, Φ_2 in $\mathfrak{H}_2$. The range of F is all of $\mathfrak{H}$. Moreover, F has a bounded inverse B with bound 1 defined in $\mathfrak{H}$ with the range $\mathfrak{H}_2$. The relations

$$FB\Phi = \Phi \ , \quad BF\Phi_2 = \Phi_2 \tag{29.2}$$

hold for all Φ in $\mathfrak{H}$, Φ_2 in $\mathfrak{H}_2$. The operators B in $\mathfrak{H}$ and F in $\mathfrak{H}_2$ are strictly self-adjoint.

This theorem will be one of our major tools in establishing strictly self-adjoint operators. To prove this theorem we first note that the bilinear form (Φ_1', Φ_1) is bounded in $\mathfrak{H}_1$ with respect to $|| \ ||_1$; in fact,

$$|\Phi_1', \Phi_1| \leq ||\Phi_1'|| \ ||\Phi_1|| \leq ||\Phi_1'||_1 \ ||\Phi||_1 .$$

Consequently, according to the corollary of Theorem 18.1, there is an operator B defined in $\mathfrak{H}_1$ with the $|| \ ||_1$ -bound 1, such that

$$(\Phi_1', \Phi_1) = (\Phi_1', B\Phi_1)_1 \quad \text{for all} \quad \Phi_1, \Phi_1' \quad \text{in} \quad \mathfrak{H}_1 .$$

Moreover,

$$||B\Phi_1||^2 \leq ||B\Phi_1||_1^2 = (B\Phi_1, \Phi_1) \leq ||B\Phi_1|| \ ||\Phi_1|| ,$$

whence $||B\Phi_1|| \leq ||\Phi_1||$. That is, B has even the bound 1 with respect to the original norm. Since the space $\mathfrak{H}_1$ was assumed to be dense in $\mathfrak{H}$, the operator B can be extended to all of $\mathfrak{H}$ as an operator with the bound 1. From $||B\Phi_1||_1^2 \leq ||B\Phi_1|| \ ||\Phi_1||$ and $||B\Phi_1|| \leq ||\Phi_1||$ we deduce $||B\Phi_1||_1 < ||\Phi_1||$ and conclude from the completeness of $\mathfrak{H}_1$ with respect to $|| \ ||_1$ that $B\Phi$ lies in $\mathfrak{H}_1$ for any Φ in $\mathfrak{H}$. The formula

$$(\Phi_1', \Phi) = (\Phi_1', B\Phi)_1 \tag{29.3}$$

therefore holds for all Φ in $\mathfrak{H}$, Φ_1' in $\mathfrak{H}_1$. Moreover, for all Φ in $\mathfrak{H}$

$$||B\Phi|| \leq ||B\Phi||_1 \leq ||\Phi|| . \tag{29.4}$$

Suppose $B\Phi = 0$. Then formula (29.3) gives $(\Phi_1', \Phi) = 0$ for all Φ_1' in $\mathfrak{H}_1$ and hence $\Phi = 0$ since $\mathfrak{H}_1$ is dense in $\mathfrak{H}$. Consequently, the operator B has a linear inverse, which we denote by

F, defined in the range of B, which we denote by $\mathfrak{H}_2$.

We know that this space $\mathfrak{H}_2$ is contained in $\mathfrak{H}_1$; we now maintain that it is dense in $\mathfrak{H}_1$ with respect to $|| \ ||_1$. For, else there would be a vector Φ_1 in $\mathfrak{H}_1$ such that $(\Phi_1, B\Phi')_1 = 0$ for all Φ' in $\mathfrak{H}$. That is to say, by (29.3) we would have $(\Phi_1, \Phi') = 0$ for all Φ' in $\mathfrak{H}$ and hence $\Phi_1 = 0$.

The relation $FB\Phi = \Phi$ follows from the definition of F; since by definition every Φ_2 in $\mathfrak{H}_2$ is of the form $\Phi_1 = B\Phi$ we have $BF\Phi_2 = BFB\Phi = B\Phi = \Phi_2$.

The formal self-adjointness of the operator B in $\mathfrak{H}_1$ follows from the relation $(B\Phi_1', \Phi_1) = (B\Phi_1', B\Phi_1)_1$ by interchanging Φ_1' and Φ_1. Since $\mathfrak{H}_1$ is dense in $\mathfrak{H}$ it follows that B is formally self-adjoint in $\mathfrak{H}$. Since B is bounded, it is strictly self-adjoint there, according to a remark made in Section 28.

Relation $(B\Phi', \Phi) = (\Phi', B\Phi)$ for all Φ, Φ' in $\mathfrak{H}$ implies $(\Phi_2', F\Phi_2) = (F\Phi_2', \Phi_2)$ for all Φ_2, Φ_2' in $\mathfrak{H}_2$. Hence F is formally self-adjoint. Suppose Φ_2 and Φ_0 in $\mathfrak{H}$ are such that

$$(\Phi_2', \Phi_0) = (F\Phi_2', \Phi_2)$$

for all Φ_2' in $\mathfrak{H}_2$; then

$$(B\Phi', \Phi_0) = (\Phi', \Phi_2)$$

for all Φ' in $\mathfrak{H}$. From the strict self-adjointness of B we conclude that $B\Phi_0 = \Phi_2$; hence Φ_2 is in $\mathfrak{H}_2$ and $F\Phi_2 = \Phi_0$. Thus the strict self-adjointness of F is proved.

In employing this theorem we shall use

<u>Remark</u> 1. Let the operator A with domain $\mathfrak{H}_A \subset \mathfrak{H}$ and range in $\mathfrak{G}$ be closed in its domain. Then the sub-inner-product

$$(\Phi', \Phi)_1 = (A\Phi', A\Phi) + (\Phi', \Phi)$$

is closed in $\mathfrak{H}_A$. A form $(A\Phi',A\Phi)$ itself is closed if the form $c(A\Phi',A\Phi)$ is a sub-inner-product for a suitable $c > 0$; i.e., if $c||A\Phi|| \geq ||\Phi||$. In that case the operator A is called inverse-bounded.

Clearly, if Φ^ν is a Cauchy sequence with respect to the sub-norm $|| \; ||_1$, it is a Cauchy sequence with respect to the norm $|| \; ||$ of the space $\mathfrak{H}$. Consequently, there are vectors Φ in $\mathfrak{H}$, Φ_1 in $\mathfrak{G}$ such that $\Phi^\nu \to \Phi$, $A\Phi^\nu \to \Phi_1$. The closedness of A now implies that Φ is in $\mathfrak{H}_A$ and $A\Phi = \Phi_1$; and this fact just expresses the completeness of the space $\mathfrak{H}_A$ with respect to the subnorm.

We may now formulate

Theorem 29.2. Let the operator A be closed in its domain $\mathfrak{H}_A$ and let $\mathfrak{H}_A$ be dense in $\mathfrak{H}$. Let A^* in $\mathfrak{G}_{A^*} \subset \mathfrak{G}$ be the strict adjoint of A in $\mathfrak{H}$. Then the subspace $\mathfrak{H}_{A^*A}$ of all Φ in $\mathfrak{H}_A$ for which $A\Phi$ lies in $\mathfrak{G}_{A^*}$ is dense in $\mathfrak{H}_A$. The operator A^*A is strictly self-adjoint in $\mathfrak{H}_{A^*A}$. Moreover, the range of the operator A^*A+1 is the full space $\mathfrak{H}$. The operator A^*A+1 has a bounded inverse (with bound 1), say $B = (A^*A+1)^{-1}$, and the range of B is $\mathfrak{H}_{A^*A}$, i.e.:

$$(A^*A+1)B = 1 \;, \quad B(A^*A+1) = 1 \quad \text{in} \quad \mathfrak{H}_{A^*A} \;.$$

In view of Remark 1 made above it is clear that Theorem 29.1 applies. A bounded selfadjoint operator B exists with range $\mathfrak{H}_F$ in $\mathfrak{H}_A$ such that relation

$$(*) \qquad (\Phi',\Phi) = (A\Phi',AB\Phi) + (\Phi',B\Phi)$$

holds for all Φ in $\mathfrak{H}$, Φ' in $\mathfrak{H}_A$. In $\mathfrak{H}_F$ a strict self-adjoint inverse F of B is defined with range $\mathfrak{H}$. We must show that $\mathfrak{H}_F = \mathfrak{H}_{A^*A}$.

Here we make use of the assumption that A in $\mathfrak{H}_A$ possesses a strict adjoint A^* in $\mathfrak{G}_{A^*}$. From the second property of strict adjointness with $AB\Phi$ and $\Phi-B\Phi$ instead of Φ and Φ_1 we conclude from (*) that $AB\Phi$ is in $\mathfrak{H}_{A*}$ and that $A*AB\Phi = (1-B)\Phi$; i.e., $B\Phi$ is in $\mathfrak{H}_{A*A}$ and $(A*A+1)B\Phi = \Phi$. Conversely, if Φ' is in $\mathfrak{H}_{A*A}$ we may deduce from (*) the relation $(\Phi'-B(A*A+1)\Phi',\Phi) = 0$ whence $B(A*A+1)\Phi' = \Phi'$; hence Φ' is in $\mathfrak{H}_F$. The statement then follows.

In addition to the operator $B = (A*A+1)^{-1}$ the operator

$$C = A(A*A+1)^{-1}$$

plays a role; it is defined since the range of B is in $\mathfrak{H}_A$. Moreover, C is bounded with bound 1; for we have

$$||C\Phi||^2 = ||AB\Phi||^2 \le ||B\Phi||_1^2 \le ||\Phi||^2 .$$

Suppose now the domain $\mathfrak{E}_{A*}$ of the operator $A*$ is dense in $\mathfrak{E}$ so that $A*$ has an adjoint $A**$. Then Theorem 29.2 can be applied to $A*$ and an operator $A**A*$ can be established. Now by virtue of V. Neumann's Theorem 28.4 we have $A** = A$, hence $A**A* = AA*$. With this in mind we formulate

<u>Theorem</u> 29.3. Suppose the operators A, $A*$ are strictly adjoint to each other in domains $\mathfrak{H}_A$, $\mathfrak{E}_{A*}$ dense in $\mathfrak{H}$ and $\mathfrak{E}$. Then the operator $(AA*+1)^{-1}A$ defined in $\mathfrak{H}_A$ equals C:

$$(AA*+1)^{-1}A = A(A*A+1)^{-1} = C \quad \text{in} \quad \mathfrak{H}_A .$$

To prove it let Φ be in $\mathfrak{H}_A$ and set $\Phi_1 = B\Phi$ so that $A*A\Phi_1 = \Phi-\Phi_1$. It follows that $A*A\Phi_1$ is in $\mathfrak{H}_A$ and $AA*A\Phi_1 = A\Phi-A\Phi_1$, or $(AA*+1)A(A*A+1)^{-1}\Phi = A\Phi$ whence the statement after division by $(A*A+1)$, i.e. after multiplication by $B = (A*A+1)^{-1}$.

Theorems 29.2 and 29.3 will be used in the next sections. Theorem 29.2 and other applications of Theorem 29.1 will be used in our treatment of differential operators.

Statements analogous to Theorems 29.2 and 29.3 could be proved - by essentially the same arguments - in the case in which $c||A\Phi|| \geq ||\Phi||$. We do not carry this out.

The case in which the closed operator A is formally self-adjoint, $A^{\dagger} = A$, and $\mathfrak{C} = \mathfrak{H}$, is of particular importance. In this case the form $(A\Phi, A\Phi)$ can be written as $(\Phi, A^2\Phi)$ in a subspace $\mathfrak{C}_{A^2}$ of the domain $\mathfrak{C}_A$. One may wonder whether or not Theorem 29.2 yields the inverse of A^2+1 (or of A^2) in this case. That need not be so; for it may happen that the formal adjoint $A^{\dagger} = A$ is not the strict adjoint A^* of A. In fact, the strict adjoint A^* need not even be formally self-adjoint in such a case. Examples of this occurrence will be given in connection with differential operators in a later section.

The question naturally arises whether or not a closed formally self-adjoint operator can be extended to a strictly self-adjoint one. The domain of this extension would then be a proper subspace of the domain of A^*, unless A^* itself is strictly self-adjoint.

Such an extension can be constructed if the operator A is inverse-bounded. This extension will at first be carried out not for the operator A but for the operator $E = A^2$. The extension of A can then be obtained from that of E as $A = E^{1/2}$ once the functional calculus for E has been established. We shall formulate and prove the pertinent extension theorem at the end of Section 31.

It should be mentioned that, once an inverse B of $E = A^2$, and hence the inverse BA of A, has been found, the spectral resolution of A can be derived directly from those of B and BA, without employing the theory of Section 30 to be developed. We do not carry this out.

30. Spectral Resolution of Self-Adjoint Operators

The operator A defined in a dense subspace $\mathfrak{H}_A$ of $\mathfrak{H}$ is strictly self-adjoint, or "self-adjoint" for short, if firstly relation

(1) $(\Phi',A\Phi) = (A\Phi',\Phi)$ holds for all Φ,Φ' in $\mathfrak{H}_A$,

and if second a vector Φ in $\mathfrak{H}$ belongs to $\mathfrak{H}_A$ with $A\Phi = \Phi_1$ whenever the relation

(2) $(\Phi',\Phi_1) = (A\Phi',\Phi)$ holds for all Φ in $\mathfrak{H}_A$.

We shall establish a functional calculus for such operators using a variant of the original method of Neumann. We observe that the hypothesis of Theorem 29.2 is satisfied for this operator A with $A^* = A$ and $\mathfrak{C} = \mathfrak{H}$. Consequently there exists a bounded self-adjoint operator $B = (A^2+1)^{-1}$ with bound 1 whose range $\mathfrak{H}_{A^2}$ consists of all Φ in $\mathfrak{H}_A$ for which $A\Phi$ is in $\mathfrak{H}_A$. Also, the operator $C = A(A^2+1)^{-1}$ is defined in $\mathfrak{H}$ and bounded with the bound 1.

We now maintain that these operators B and C commute:

$$BC = CB .$$

This follows immediately from Theorem 29.3, according to which

$$(A^2+1)^{-1}A(A^2+1)^{-1} = A(A^2+1)^{-1}(A^2+1)^{-1} .$$

Consequently (as was already mentioned in the last section, 23, of Chapter IV), all piecewise continuous functions of B and C commute. With the aid of such functions of B and C we shall define functions of A.

First we restrict ourselves to piecewise continuous bounded functions $f(\alpha)$ of α which are continuous for $|\alpha| \geq a$ with a suitable $a > 0$ and approach definite limits as $\alpha \to \pm\infty$. Every such function can be written as the sum of an even and an odd function of α. Every even function of α may be regarded as a function $g(\beta)$ of

$$\beta = \frac{1}{\alpha^2+1} ,$$

while each odd function of α is of the form $\theta(\gamma)h(\gamma)$, where

$$\gamma = \frac{\alpha}{\alpha^2+1}$$

and

$$\theta(\gamma) = 1 \text{ for } \gamma \geq 0, \quad = -1 \text{ for } \gamma < 0 ;$$

note that $\theta^2(\gamma) = 1$ for all γ.

We therefore can write $f(\alpha)$ uniquely in the form

$$f(\alpha) = g(\beta) + \theta(\gamma)h(\beta) .$$

where

$$\begin{aligned} g(\beta) \pm h(\beta) &= f(\pm\sqrt{\beta^{-1} - 1}) && \text{for } 0 < \beta \leq 1 , \\ &= f(0) && \text{for } \beta \geq 1 , \\ &= 0 && \text{for } \beta \leq 0 . \end{aligned}$$

By virtue of the assumptions made on $f(\alpha)$, the functions g and h are piecewise continuous and bounded. We can therefore define the operator $f(A)$ by

$$f(A) = g(B) + \theta(C)h(B) . \qquad (3.0.1)_A$$

Since $\theta(\gamma)$ is real the operator $\theta(C)$ is self-adjoint; if $f(\alpha)$ is real, the same is true of g and h. Hence $g(B)$, $h(B)$ and $f(A)$ are self-adjoint. Note that the operator $f(A)$ is bounded since $f(\alpha)$ is.

In order to show that the rules of operational calculus are obeyed, let

$$f_1(\alpha)f_2(\alpha) = f(\alpha) ;$$

then, evidently,

$$g(\beta) = g_1(\beta)g_2(\beta) + h_1(\beta)h_2(\beta) ,$$

$$h(\beta) = g_1(\beta)h_2(\beta) + h_1(\beta)g_2(\beta) .$$

From the fact that functions of B and C commute we conclude

$$g(B) = g_1(B)g_2(B) + h_1(B)h_2(B) ,$$

$$h(B) = g_1(B)h_2(B) + h_1(B)g_2(B) ,$$

and hence

$$f_1(A)f_2(A) = f(A) .$$

We need not prove the rather obvious addition rule.

Next we must extend the definition of $f(A)$ to piecewise continuous not necessarily bounded functions. To any such function $f(\alpha)$ we introduce the "cut-off" function

$$f_a(\alpha) = f(\alpha) \quad \text{for} \quad |\alpha| \leq a, \quad = 0 \text{ for } \quad |\alpha| \geq a .$$

Suppose now that the vector Φ is such that the norms $||f_a(A)\Phi||$ remain bounded as $a \to \infty$. Then, we maintain, $f_a(A)\Phi$ converges to a limit vector, which we denote by $f(A)\Phi$. Clearly, $[f_{a'}(\alpha)-f_a(\alpha)]^2 = f_{a'}^2(\alpha)-f_a^2(\alpha)$ for $a' > a$. Hence $[f_{a'}(A)-f_a(A)]^2 = f_{a'}^2(A)-f_a^2(A)$ and

$$||[f_{a'}(A)-f_a(A)]\Phi||^2 = ||f_{a'}(A)\Phi||^2 - ||f_a(A)\Phi||^2 \to 0$$

since $||f_a(A)\Phi||$ is monotonic as $a \to \infty$. Thus, $f_a(A)\Phi$ is a Cauchy sequence. Obviously, the operator $f(A)$ defined by the limit is linear. The space $\mathfrak{H}_{f(A)}$ of all vectors Φ for which $f(A)$ is defined in this manner is evidently dense in $\mathfrak{H}$. Moreover, as we shall show, the operator $f(A)$ is self-adjoint in $\mathfrak{H}_{f(A)}$.

The formal self-adjointness of $f(A)$ follows from the fact

that the approximating operators have this property. To prove the strict self-adjointness, let Φ and Φ_1 in $\mathfrak{H}$ be such that

$$(\Phi',\Phi_1) = (f(A)\Phi',\Phi)$$

holds for all Φ' in $\mathfrak{H}_{f(A)}$. Then this relation holds for all $\Phi'_a = \eta_a(A)\Phi'$ where $\eta_a(\alpha)$ is the unit step function of $|\alpha| \leq a$. Now, by definition of $f(A)$ we have $f(A)\eta_a(A) = f_a(A)$; hence the above relation for Φ'_a instead of Φ', i.e. $(\eta_a(A)\Phi',\Phi_1) = (f_a(A)\Phi',\Phi_1)$, yields

$$(\Phi',\eta_a(A)\Phi_1) = (\Phi',f_a(A)\Phi)$$

for all Φ' in $\mathfrak{H}_{f(A)}$, hence for all Φ' in $\mathfrak{H}$. Therefore

$$f_a(A)\Phi = \eta_a(A)\Phi_1 \rightarrow \Phi_1 \quad \text{as} \quad a \rightarrow \infty .$$

By definition, then, Φ is in $\mathfrak{H}_{f(A)}$ and $f(A)\Phi = \Phi_1$.

Note that the operator A itself is one of those functions of A now defined; one readily verifies that the new operator A and its domain agree with the old one. Thus we have established all essential features of the functional calculus for the operator A. In particular, the spectral projectors $\eta_{\mathscr{I}}(A)$ and the eigenspaces into which they project are defined.

We add a few remarks about the spectral representation of the self-adjoint operator A. We confine ourselves to considering a space $\mathfrak{C}(\Phi_0)$ which is generated from a vector Φ_0 in $\mathfrak{H}$ by all bounded piecewise continuous functions $f(A)$; i.e., which is the closure of the space of all $f(A)\Phi_0$; see Section 22.

Let $\mathscr{I}$ be any bounded α-interval with the unit step function $\eta_{\mathscr{I}}(\alpha)$. Then we may define a measure function $r(\alpha)$ such that

$$\Delta_{\mathscr{I}} r = ||\eta_{\mathscr{I}}(A)\Phi_0||^2$$

in obvious notation. Let $f(\alpha)$ be any piecewise continuous complex-valued function and $f_a(\alpha)$ its cut-off function. From the spectral representation of bounded operators, Section 22, we deduce the relations

$$||f_a(A)\Phi_0||^2 = \int_{|\alpha|\leq a} |f(\alpha)|^2 dr(\alpha)$$

and

$$||f_{a'}(A)\Phi_0 - f_a(A)\Phi_0||^2 = \int_{a<|\alpha|\leq a'} |f(\alpha)|^2 dr(\alpha) .$$

The condition that Φ_0 is in the domain of $f(A)$ then is exactly given by the condition

$$(*) \qquad \int_{-\infty}^{\infty} |f(\alpha)|^2 dr(\alpha) < \infty .$$

The vectors Φ of the space $\mathfrak{G}(\Phi_0)$ are now represented by the (ideal) functions $f(\alpha)$ obtained by closure from the piecewise continuous functions $f(\alpha)$ satisfying (*) :

$$\Phi \Longleftrightarrow f(\alpha) .$$

The functions $f(\alpha)$ which together with the function $f_1(\alpha) = f(\alpha)$ satisfy condition (*) clearly represent those vectors $\Phi = f(A)\Phi_0$ for which also the vector $\Phi_1 = f_1(A)\Phi_0 = Af(a)\Phi_0$ belongs to $\mathfrak{G}(\Phi_0)$. In other words, these functions represent the domain of A. Thus we see that the domain of the operator A is represented by those functions $f(\alpha)$ for which

$$\int \alpha^2 |f(\alpha)|^2 dr(\alpha) < \infty .$$

This result implies the fact that <u>the domain of an unbounded</u>

self-adjoint operator cannot be the whole Hilbert space. For, if $r(\alpha)$ is constant for $\alpha > \alpha_0$ and for $\alpha < -\alpha_0$ for sufficiently large α the operator A is bounded with the bound α_0. If, on the other hand, the carrier of $r(\alpha)$ is not bounded there evidently exist functions $f(\alpha)$ with

$$\int |f(\alpha)|^2 dr(\alpha) < \infty \quad \text{but} \int \alpha^2 |f(\alpha)|^2 dr(\alpha) = \infty .$$

Clearly, i.e., there are vectors in the Hilbert space which are not in the domain of the operator A.

31. Closeable Forms

In Section 29 we introduced the notion of closedness of a bilinear form. We shall now slightly extend this notion and call a (Hermitean) bilinear non-negative form $(\Phi',\Phi)'$ closed in its domain $\mathfrak{H}_1$ if the relations $||\Phi^\nu-\Phi^\mu||' \to 0$ for Φ^ν,Φ^μ in $\mathfrak{H}_1$ - in obvious notation - and $||\phi^\nu-\Phi|| \to 0$ for a Φ in $\mathfrak{H}$ imply that Φ is in $\mathfrak{H}_1$ and $||\Phi^\nu-\Phi||' \to 0$. If the form $(\Phi',\Phi)'$ is closed in $\mathfrak{H}_1$, this space is obviously complete with respect to the subnorm $|| \; ||_1$ given by

$$(\Phi,\Phi)_1 = (\Phi,\Phi)' + (\Phi,\Phi) .$$

In case there is a constant $c > 0$ such that

$$c(\Phi,\Phi)' \geq (\Phi,\Phi)$$

we may take $c(\Phi,\Phi)'$ itself as subnorm $(\Phi,\Phi)_1$.

If the domain is not complete with respect to the subnorm $|| \; ||_1$, it may be possible to extend it to one that is complete. We call a form $(,)'$ defined in a domain $\mathfrak{H}'$ "closeable" if the relations

$$||\Phi^\nu-\Phi^\mu||' \to 0 \quad \text{and} \quad ||\Phi^\nu|| \to 0 \quad \text{for } \Phi^\nu,\Phi^\mu \text{ in } \mathfrak{H}'$$

imply

$$||\Phi||' \to 0 \quad \text{as} \quad \nu \to \infty .$$

If a form in $\mathfrak{H}'$ is closeable we may extend $\mathfrak{H}'$ to the space $\mathfrak{H}_1$ of all those Φ in $\mathfrak{H}$ to which there is a sequence Φ^ν in $\mathfrak{H}'$ such that

$$||\Phi^\nu - \Phi^\mu||' \to 0 \quad \text{and} \quad ||\Phi^\nu - \Phi|| \to 0 .$$

By virtue of the first relation, the inner product $(\Phi_1^\mu, \Phi_2^\nu)'$ tends to a limit. We must show that this limit depends only on the limits Φ_1, Φ_2 of the approximating sequences Φ_1^μ, Φ_2^ν and not on the choice of these sequences. To prove this we introduce differences $\tilde{\Phi}_1^\mu$, $\tilde{\Phi}_2^\nu$ of two such pairs of sequences and observe that $||\tilde{\Phi}_1^\mu|| \to 0$ and $||\tilde{\Phi}_2^\nu|| \to 0$ would imply $||\tilde{\Phi}_1^\mu||' \to 0$ and $||\tilde{\Phi}_2^\nu||' \to 0$ by virtue of the closeability assumption, so that also $(\tilde{\Phi}_1^\mu, \tilde{\Phi}_2^\nu)'$ depends only on Φ_1 and Φ_2 and moreover is bilinear. Clearly, the closedness of the form $(\Phi', \Phi)'$ in $\mathfrak{H}_1$ is now an immediate consequence. We formulate this result simply in

<u>Theorem</u> 31.1. <u>A closeable form can be extended to a closed form</u>.

One might have extended the space $\mathfrak{H}'$ to a space $\mathfrak{H}_1$ complete with respect to the associated subnorm by simply adjoining ideal elements; but then it would have been necessary to identify these ideal elements with vectors in $\mathfrak{H}$. This would not have been possible if there existed a sequence Φ^ν in $\mathfrak{H}'$ for which $||\Phi^\nu - \Phi^\mu||_1 \to 0$ and $||\Phi^\nu|| \to 0$ but not $||\Phi^\nu||_1 \to 0$. For then the ideal element would correspond to the vector $\Phi = 0$ in $\mathfrak{H}$ without being 0 in $\mathfrak{H}_1$. To exclude this occurrence, the "closeability" requirement was introduced.

Clearly, the form $(\Phi', \Phi)' = (A\Phi', A\Phi)$ is closeable in $\mathfrak{H}_A$ if the operator A in $\mathfrak{H}_A$ possesses a closure. The following theorem describes another important class of closeable operators.

<u>Theorem</u> 31.2. Let the non-negative Hermitean quadratic form $(\Phi,\Phi)'$ be the least upper bound of a set of bounded forms in a domain $\mathfrak{H}'$. Then the form $(\Phi',\Phi)'$ in $\mathfrak{H}'$ is closeable.

To prove this statement find for every $\varepsilon > 0$ a ν_ε such that $||\Phi^\nu-\Phi^\mu||' \leq \varepsilon$ for $\mu \geq \nu \geq \nu_\varepsilon$. Then also $||\Phi^\nu-\Phi^\mu||^b \leq \varepsilon$ where the superscript b refers to any of the bounded forms entering the definition of $||\ ||'$. Now, let μ tend to infinity. From the hypothesis $||\Phi^\nu|| \to 0$ and the boundedness of $||\ ||^b$ we conclude $||\Phi^\nu-\Phi^\mu||^b \to ||\Phi^\nu||^b$ as $\mu \to \infty$. Hence $||\Phi^\nu||^b \leq \varepsilon$. From the definition of $||\ ||'$ we then infer $||\Phi^\nu||' \leq \varepsilon$. Thus Theorem 31.2 is proved.

In employing Theorem 31.2 we shall make use of the obvious

<u>Lemma</u>. If the non-negative forms $(,)'$ and $(,)''$ are closeable in a common domain then so is $(,)' + (,)''$.

It follows from this lemma that the form

$$(A\Phi',A\Phi) + (\Phi',\Phi)'$$

is closeable in a domain $\mathfrak{H}'$ if the form $(\Phi',\Phi)'$ is closeable there and if the operator A defined in $\mathfrak{H}'$ possesses a closure. It furthermore follows that the sub-inner-product

$$(\Phi',\Phi)_1 = (\Phi',\Phi) + (A\Phi',A\Phi) + (\Phi',\Phi)'$$

in $\mathfrak{H}$ can be extended to a closed sub-unit-form. By virtue of Theorem 29.1 this sub-unit-form is associated with a strictly self-adjoint operator.

This procedure of defining self-adjoint operators will prove to be very helpful in establishing the strictly self-adjoint character of various differential operators.

Finally, we shall prove the extension theorem mentioned at the end of Section 29.

Extension Theorem* for inverse bounded self-adjoint operators. Any formally self-adjoint operator E defined in a domain $\mathfrak{D} = \mathfrak{D}_E$ dense in $\mathfrak{H}$, and having a lower bound c, admits an extension in a domain $\mathfrak{D}_2 \supset \mathfrak{D}$ in which it has the same lower bound as E and is strictly self-adjoint.

To prove the statement we assume the operator E satisfies the inequality

$$c(\Phi,E\Phi) \geq (\Phi,\Phi) \quad \text{for all} \quad \Phi \quad \text{in} \quad \mathfrak{H}_E \quad \text{with a constant} \quad c > 0 ,$$

and show that the form $(\Phi,E\Phi)$ is closeable.

To this end we consider a sequence $\{\Phi^\nu\}$ of vectors in $\mathfrak{D}_E$ for which

$$((\Phi^\mu-\Phi^\nu),E(\Phi^\mu-\Phi^\nu)) \to 0 \quad \text{as} \quad \mu,\nu \to \infty ;$$

while at the same time

$$||\Phi^\mu|| \to 0 \quad \text{as} \quad \mu \to \infty .$$

After writing

$$(\Phi^\nu,E\Phi^\nu) = (\Phi^\nu,E(\Phi^\nu-\Phi^\mu)) + (E\Phi^\nu,\Phi^\mu)$$

and then estimating

$$(\Phi^\nu,E\Phi^\mu) \leq (\Phi^\nu,E\Phi^\nu)^{1/2}((\Phi^\nu-\Phi^\mu),E(\Phi^\nu-\Phi^\mu))^{1/2} + ||E\Phi^\nu|| \; ||\Phi^\mu|| ,$$

we can choose ν such that

*Proved by the author in 1931.

$$(\Phi^\nu-\Phi^\mu),E(\Phi^\nu-\Phi^\mu)) < \varepsilon^2 \quad \text{for} \quad \mu \geq \nu$$

and then choose $\mu \geq \nu$ such that

$$||E\Phi^\nu|| \; ||\Phi^\mu|| < \varepsilon^2 .$$

We are then led to the inequality

$$(\Phi^\nu,E\Phi^\nu) \leq (\Phi^\nu,E\Phi^\nu)^{1/2}\varepsilon + \varepsilon^2 ,$$

which implies

$$(\Phi^\nu,E\Phi^\nu) \leq 3\varepsilon^2 .$$

That is to say, $(\Phi^\nu,E\Phi^\nu) \to 0$ as $\nu \to \infty$.

Thus we have proved the closeability of the form $c(\Phi,E\Phi)$, and it follows from Theorem 31.1 that this form can be extended to a closed form $(\Phi,\Phi)_1$ in a domain $\mathfrak{D}_1 \supset \mathfrak{D}_E$ for which $(\Phi,\Phi)_1 \geq (\Phi,\Phi)$ holds.

Denoting the extension of $c(\Phi',E\Phi)$ by $(\Phi',\Phi)_1$, we can apply Theorem 31.2. It yields the existence of a self-adjoint operator F in a domain $\mathfrak{D}_2 \supset \mathfrak{D}_1$ such that

$$(\Phi_1',F\Phi_2) = (\Phi_1',\Phi_2)_1$$

holds for all Φ_1' in $\mathfrak{D}_1$, Φ_2 in $\mathfrak{D}_2$. Hence

$$(\Phi_1',F\Phi_2) = c(E\Phi_1',\Phi_2)$$

holds for all Φ_1' in $\mathfrak{D}_E$ and all Φ_2 in $\mathfrak{D}_2$. From the self-adjointness of F in $\mathfrak{D}_2$ it then follows that Φ_1' is in $\mathfrak{D}_2$ and $cE\Phi_1' = F\Phi_1'$. That is to say, $c^{-1}F$ in $\mathfrak{D}_2$ is a self-adjoint extension of E in $\mathfrak{D}_F$. Clearly

$$c(\Phi, c^{-1}F\Phi) = (\Phi,\Phi)_1 \geq (\Phi,\Phi)$$

holds for all Φ in $\mathfrak{D}_2$. Thus our statement has been proved.

CHAPTER VII

DIFFERENTIAL OPERATORS

32. Regular Differential Operators

In Section 10 we introduced the differential operator D acting on functions of a real variable running from $-\infty$ to $+\infty$ which have a piecewise continuous derivative. The independent variable will now be denoted by $\phi(x)$ so that the operator D transforms $\phi(x)$ into $D\phi(x) = \frac{d}{dx}\phi(x)$.

In this section we shall show that this operator D can be extended into a dense subspace $\mathfrak{D}$ of $\mathfrak{H}$ where it is closed and where it is strictly adjoint to the operator $-D$, so that the operator iD is strictly self-adjoint. In later sections we shall consider various modifications: we shall define various operators of first and second order, defined in finite or infinite domains of one or more variables and shall prove that these operators are strictly self-adjoint.

The space of functions $\phi(x)$ with a piecewise continuous derivative - for which the operator D was defined so far - will be denoted by $\mathfrak{C}_1'$; the space of those functions in $\mathfrak{C}_1'$ which are square-integrable and for which $D\phi(x)$ is square-integrable will be denoted by $\mathfrak{D}'$; the space of functions in $\mathfrak{D}'$ with bounded support will be denoted by $\dot{\mathfrak{D}}'$. With reference to the unit form

$$(\phi,\phi) = \int |\phi(x)|^2 dx$$

the operators D in $\mathfrak{D}'$ and $-D$ in $\dot{\mathfrak{D}}'$ are formally adjoint to each other since evidently

$$(\phi,D\phi') + (D\phi,\phi') = \int \{\overline{\phi(x)}\, D\phi'(x) + \overline{D\phi(x)}\, \phi'(x)\}dx = 0$$

for ϕ in $\mathfrak{D}'$ and ϕ' in $\mathring{\mathfrak{D}}'$.

Clearly, the domain $\dot{\mathfrak{D}}'$, and hence also $\mathfrak{D}'$, is dense in the Hilbert space $\mathfrak{H}$ since this space was defined by extension from the space of piecewise continuous functions with bounded support and this space in turn obviously contains $\dot{\mathfrak{D}}'$ densely. We therefore know from Theorem 28.3 (Chapter VI, Section 28) that the operators -D and D in $\dot{\mathfrak{D}}'$ and $\mathfrak{D}'$ possess closures in extended domains which we denote by $\dot{\mathfrak{D}}$ and $\mathfrak{D}$ respectively; the closed extension of D in these domains will also be denoted by D. Of course, the operators D and -D in $\mathfrak{D}$ and $\dot{\mathfrak{D}}$ are formally adjoint.

We now maintain that <u>these two spaces $\mathfrak{D}$ and $\dot{\mathfrak{D}}$ are the same</u>. In other words, in the definition of the space $\mathfrak{D}$ by extension of a space of smooth functions it is no restriction to require that these functions have bounded support.

<u>Theorem 32.1</u>. $\mathfrak{D} = \dot{\mathfrak{D}}$.

<u>Proof</u>. Denote by $\eta_a(x)$ and $\zeta_a(x)$ the functions defined by

$$\eta_a(x) \begin{cases} = 1 & |x| \leq a , \\ = 0 & |x| \geq a ; \end{cases}$$

$$\zeta_a(x) \begin{cases} = 1 & |x| \leq a , \\ = a+1-|x| & a \leq |x| \leq a+1 , \\ = 0 & a+1 \leq |x| . \end{cases}$$

Let $\phi(x)$ be in $\mathfrak{D}'$; then $\phi_a = \zeta_a\phi$ is in $\dot{\mathfrak{D}}'$. Note

$$D\zeta_a\phi = \zeta_a D\phi + (D\zeta_a)\phi .$$

Now, $D\zeta_a \leq 1 - \eta_a$, $1 - \zeta_a \leq 1 - \eta_a$. Hence

$$||D\phi_a - D\phi|| \le ||D\phi_a - \zeta_a D\phi|| + ||(1 - \zeta_a)D\phi||$$

$$\le ||(1-\eta_a)\phi|| + ||(1-\eta_a)D\phi|| \to 0 \quad \text{as} \quad a \to \infty$$

and at the same time $||(1-\zeta_a)\phi|| \le ||(1-\eta_a)\phi|| \to 0$.

From the definition of the space $\mathfrak{D}$ as the domain of the closure of D in $\dot{\mathfrak{D}}'$ it then follows that ϕ is in $\mathfrak{D}$. Hence $\mathfrak{D} \subset \dot{\mathfrak{D}}$ is proved.

Since the inclusion $\dot{\mathfrak{D}} \subset \mathfrak{D}$ is obvious, the statement of Theorem 32.1 is proved.

From Theorem 28.1 (Chapter VI, Section 28) we know that the operator D in $\mathfrak{D}$ possesses a strict adjoint: $-D$ in $\mathfrak{D}^*$, an extension of an extension of $-D$ in the space $\dot{\mathfrak{D}}'$ and hence just as well of the closure $-D$ in $\dot{\mathfrak{D}} = \mathfrak{D}$. We now maintain that even the space $\mathfrak{D}^*$ is the same as the space $\mathfrak{D}$.

<u>Theorem</u> 32.2. $\mathfrak{D}^* = \mathfrak{D}$.

In other words, the operator D in $\mathfrak{D}$ is strictly adjoint to $-D$ in $\mathfrak{D}$; or iD in $\mathfrak{D}$ is strictly self-adjoint there. To prove this theorem, let $\phi(x)$ be a function in $\mathfrak{D}^*$; i.e., a function in $\mathfrak{H}$ to which there is a function $\phi_1(x)$ in $\mathfrak{H}$ such that the relation

$$\int \psi(x)\phi_1(x)\,dx = -\int \overline{D\psi\ (x)}\,\phi(x)\,dx$$

holds for all $\psi(x)$ in $\mathfrak{D}$, hence in particular for all $\psi(x)$ in $\dot{\mathfrak{D}}$. To show that this $\phi(x)$ is in $\mathfrak{D}$ we must show that there exists a sequence of functions $\phi^\nu(s)$ in $\dot{\mathfrak{D}}'$ such that

$$||\phi^\nu - \phi|| \to 0\ , \quad ||D\phi^\nu - \phi_1|| \to 0 \quad \text{as } \nu \to \infty .$$

Such a sequence of approximating functions can be formed in

different ways. We shall present a method which can easily be extended to a rather large class of ordinary and partial differential operators although for ordinary differential operators one could proceed in a somewhat simpler manner.

We shall construct the functions $\phi^{\nu}(x)$ by applying certain appropriate smoothing operators on the function $\phi(x)$. These valid smoothing operators are given as integral operators

$$J^{\nu}\phi(x) = \int j^{\nu}(x-x')\phi(x')dx'$$

whose kernels are functions of $x-x'$ so chosen that

$$\int j(x-x')dx' = 1 . \tag{32.1}$$

Specifically, we choose a non-negative function $j(\xi)$ in $\mathfrak{C}_1$ supported by the interval $|\xi| \leq 1$ and for which

$$\int_{-1}^{1} j(\xi)d\xi = 1 ,$$

and then set

$$j^{\nu}(x-x') = \nu j(\nu(x-x')) .$$

Clearly, this kernel is supported by $|x-x'| \leq 1/\nu$ and satisfies condition (32.1). If $\dot{\phi}$ is a function in $\mathfrak{C}_1'$ with bounded support in $|x| \leq a$, say, the function $J^{\nu}\dot{\phi}(x)$ is defined and has also bounded support, namely in $|x| \leq a + 1/\nu$. Since the Holmgren norm of the operator J^{ν} evidently equals 1 (see Chapter IV, Section 20), the inequality

$$||J^{\nu}\phi|| \leq ||\phi|| \tag{32.2}$$

holds for such functions. This inequality shows that the operator

J^ν can be extended to all of $\mathfrak{H}$.

Operators of this type were treated at the end of Section 20 in Chapter IV. There we saw that the operators J^ν approximate the identity strongly. I.e., for every function $\phi(x)$ in $\mathfrak{H}$ the functions $J^\nu\phi(x)$ tend to $\phi(x)$; i.e.,

$$||J^\nu\phi-\phi|| \to 0 \quad \text{as} \quad \nu \to \infty. \tag{32.3}$$

Integral operators of the form J^ν which approximate the identity strongly have been called "mollifiers". The reason is that they transform the function $\phi(x)$ into a function $\phi^\nu(x) = J^\nu\phi(x)$ which is close to $\phi(x)$ in the sense of the norm $||\ ||$, but are smoother than the function $\phi(x)$ is required to be. The degree of smoothing depends on the choice of the kernel j^ν. We have required this kernel to have continuous derivatives. As a consequence the functions $J^\nu\phi(x)$ have continuous derivatives, as we proceed to show.

For any finite interval $|x| \leq x_0$ the inequality

$$\max_{|x|\leq x_0} |J^\nu\phi(x)| \leq C_\nu||\phi||$$

holds with an appropriate constant C_ν. For this relation evidently holds for any $\phi = \dot{\phi}$ in $\mathfrak{C}$. From this inequality we conclude that $J^\nu\phi(x)$ is approximated uniformly by $J^\nu\dot{\phi}(x)$ in $|x| \leq x_0$ if ϕ in $\mathfrak{H}$ is approximated with respect to $||\ ||$ by $\dot{\phi}$ in $\mathring{\mathfrak{C}}$. Consequently, $J^\nu\phi(x)$ <u>is continuous for</u> ϕ <u>in</u> $\mathring{\mathfrak{C}}$. Moreover, the function $J^\nu\phi(x)$ has a continuous derivative $DJ^\nu\phi(x)$ if $\phi(x)$ is in $\dot{\mathfrak{C}}$. Since for such functions the inequality

$$\max_{|x|\leq a_0} |DJ^\nu\phi(x)| \leq C'_\nu||\phi||$$

holds with an appropriate constant C'_ν, it follows that the function $J^\nu\phi(x)$ <u>has a continuous derivative for any function</u> ϕ <u>in</u> $\mathfrak{H}$.

We are now ready to prove the statement $\mathfrak{D}^* = \mathfrak{D}$ made above. Let $\phi(x)$ be in $\mathfrak{D}^*$; then we know that $\phi^\nu(x) = J^\nu\phi(x)$ is in $\mathfrak{C}'$, and that $||\phi^\nu - \phi|| \to 0$. Moreover,

$$DJ^\nu\phi(x) = \int Dj^\nu(x-x')\phi(x')dx' = -\int D'j^\nu(x-x')\phi(x')dx'$$

where $D' = d/dx'$. We now make use of the assumption that ϕ is in $\mathfrak{D}^*$ so that there is a ϕ_1 in $\mathfrak{H}$ such that

$$\int \tilde{\phi}(x')\phi_1(x')dx' = -\int D'\tilde{\phi}(x')\phi(x')dx'$$

for all $\tilde{\phi}$ in $\mathfrak{D}$, in particular for all $\tilde{\phi}$ in $\mathfrak{D}'$. Now, for each point x chosen the kernel $j^\nu(x-x')$ considered as a function of x' is such a function in $\mathring{\mathfrak{D}}$ since it vanishes for $|x'-x| \geq 1/\nu$. Consequently, we may conclude that

$$\int D'j^\nu(x-x')\phi(x')dx' = -\int j^\nu(x-x')\phi_1(x')dx$$

$$= -J^\nu\phi_1(x) \ .$$

In other words, we have

$$DJ^\nu\phi(x) = J^\nu\phi_1(x) \ .$$

From $||J^\nu\phi_1 - \phi_1|| \to 0$ we therefore may conclude

$$||DJ^\nu\phi - \phi_1|| \to 0 \ .$$

Thus we have shown that ϕ is in the domain $\mathfrak{D}$ of the closure of D in $\mathfrak{D}'$; i.e., $\mathfrak{D}^* \subset \mathfrak{D}$.

Since the opposite relation $\mathfrak{D} \subset \mathfrak{D}^*$ is obvious, Theorem 32.2 is proved.

33. Ordinary Differential Operators in a Semi-Bounded Domain

In this section we shall deal with the differential operator d/dx acting on functions $\phi(x)$ defined on the half-axis $0 \leq x < \infty$. We shall carry over the method of Section 32; but in doing this we shall get noticeably different results.

The functions $\phi(x)$ defined for $0 \leq x \leq \infty$ which have piecewise continuous derivatives and for which

$$(\phi,\phi) = \int_0^\infty |\phi(x)|^2 \, dx \,, \quad (D\phi,D\phi) = \int_0^\infty |D\phi(x)|^2 dx$$

are finite are said to form the space $\mathfrak{D}'$; the space $\mathring{\mathfrak{D}}'$ consists of the functions in $\mathfrak{D}'$ with bounded support. In addition, we introduce the spaces $\mathfrak{D}_0'$ and $\mathring{\mathfrak{D}}_0'$ of those functions $\phi(x)$ in $\mathfrak{D}'$ and $\dot{\mathfrak{D}}'$ for which

$$\phi(0) = 0 \,.$$

The operator $D = d/dx$, when restricted to functions in $\mathfrak{D}_0'$ or $\mathring{\mathfrak{D}}'$, will be denoted by D_0. One immediately verifies that the operator D_0 in $\mathfrak{D}_0'$ and $-D$ in $\mathfrak{D}'$ as well as D_0 in $\mathfrak{D}_0'$ and $-D$ in $\mathfrak{D}'$ are formally adjoint to each other. Since the spaces $\mathfrak{D}'$, $\mathfrak{D}'$, $\mathring{\mathfrak{D}}_0$ and $\mathfrak{D}_0'$ are obviously dense in $\mathfrak{H}$ it is clear that the operator D in $\breve{\mathfrak{D}}'$ and $\mathring{\mathfrak{D}}'$ as well as D_0 in $\mathfrak{D}_0'$ and $\mathring{\mathfrak{D}}_0'$ can be closed. The domain of these closures will be denoted by $\mathfrak{D}$, $\dot{\mathfrak{D}}$, $\mathfrak{D}_0$, $\dot{\mathfrak{D}}_0$.

In the same manner in which Theorem 1 was proved in Section 32 one may prove

<u>Theorem</u> 33.1. $\mathfrak{D} = \mathring{\mathfrak{D}}$, $\quad \mathfrak{D}_0 = \mathring{\mathfrak{D}}_0$,

so that the superscript • may be omitted. The question naturally arises whether or not the subscript can also be omitted. We shall show that this is not so.

Theorem 33.1°. $\mathfrak{D} \neq \mathfrak{D}_0$.

In other words, the boundary condition $\phi(0) = 0$ is essential. The proof of this statement is based on various lemmas.

Lemma 1. For every function $\phi(x)$ in $\mathfrak{D}'$ the inequality

$$|\phi(x)| \leq \sqrt{a}||D\phi|| + \frac{1}{\sqrt{a}}||\phi||$$

holds for every $a > 0$ and every x in $0 \leq x \leq a$.

To prove it we need only observe that by the Schwarz inequality the relation

$$|\phi(x)| = |\int_{x'}^{x} D\phi(x'')dx'' + \phi(x')| \leq \sqrt{a}\ ||D\phi|| + |\phi(x')|$$

holds for $0 \leq x \leq a$; integration with respect to x from 0 to a together with division by a then yields the statement.

Lemma 2. $\mathfrak{D} \subset \mathfrak{C}$. That is, every function in $\mathfrak{D}$ is continuous; more precisely, every (ideal) function in $\mathfrak{D}$ "equals" a continuous function.

Let ϕ be a function in $\mathfrak{D}$ and $\phi^{\nu}(x)$ be a sequence in $\mathfrak{D}'$ such that $||\phi^{\nu}-\phi|| \to 0$, $||D\phi^{\nu}-D\phi|| \to 0$ as $\nu \to 0$. Then $||\phi^{\nu}-\phi^{\mu}|| \to 0$, $||D\phi^{\nu}-D\phi^{\mu}|| \to 0$ as $\nu,\mu \to \infty$. From the inequality of Lemma 1 we may then conclude that the sequence $\phi^{\nu}(x)-\phi^{\mu}(x)$ converges uniformly to zero in every interval $0 \leq x \leq a$, so that in each such interval $\phi^{\nu}(x)$ converges uniformly to a limit function $\tilde{\phi}(x)$ which of course is continuous. Clearly $||\phi^{\nu}-\phi|| \to 0$ and the fact that ϕ^{ν} tends uniformly to $\tilde{\phi}$ in $0 \leq x \leq a$ implies $\int_0^a|\phi-\tilde{\phi}|^2dx = 0$, so that $\tilde{\phi} = \phi$.

Lemma 3. $\mathfrak{D}_0$ is the space of all $\phi(x)$ in $\mathfrak{D}$ for which $\phi(0) = 0$.

Note that this last condition makes sense since $\phi(x)$ is continuous by Lemma 2. Clearly, if the functions $\phi^{\nu}(x)$ approximating $\phi(x)$ vanish at $x = 0$ the same is true for $\phi(x)$ by Lemma 1; thus any function in $\mathfrak{D}_0$ vanishes at $x = 0$. Conversely, let $\phi(x)$ be a function in $\mathfrak{D}$ which vanishes at $x = 0$. Then there is a sequence of functions in $\mathfrak{D}'$, say $\phi^{\nu}(x)$, such that $||\phi^{\nu}-\phi|| \to 0$, and $||D\phi^{\nu}-D\phi|| \to 0$. According to Lemma 1 this implies pointwise convergence; hence $\phi^{\nu}(0) \to 0$. Now the elements of the sequence

$$\phi^{\nu'}(x) = \phi^{\nu}(x) - \phi^{\nu}(0)\, e^{-\tau^2 x}$$

are in $\mathfrak{D}_0'$. Taking $\tau = \tau_{\nu}$ such that $\tau_{\nu}|\phi^{\nu}(0)| \to 0$, we clearly have

$$||\phi^{\nu'}-\phi|| \to 0 \quad \text{and} \quad ||D\phi^{\nu'}-D\phi|| \to 0 \; ,$$

which shows that if ϕ is in $\mathfrak{D}$ and $\phi(0) = 0$ then ϕ is in $\mathfrak{D}_0$.

Corollary. Let $\phi_1(x)$ be any function in $\mathfrak{D}$ not in $\mathfrak{D}_0$. Then $\mathfrak{D}$ consists exactly of the functions $\phi(x)$ of the form

$$\phi(x) = c\phi_1(x) + \phi_0(x)$$

with $\phi_0(x)$ in $\mathfrak{D}_0$.

It follows from Lemma 3 that Theorem 33.1° is proved as soon as a single function $\phi(x)$ in $\mathfrak{D}$ with $\phi(0) \neq 0$ is exhibited. Such a function evidently exists.

Clearly, then, we must consider both spaces $\mathfrak{D}$ and $\mathfrak{D}_0$. It is obvious that the operators $-D$ and D_0 in these spaces are formally adjoint. Since these spaces are dense in $\mathfrak{H}$ the operators D and D_0 have strict adjoints D^* and D_0^*; their domains will be denoted by $\mathfrak{D}^*$ and $\mathfrak{D}_0^*$. Evidently, we have the inclusions

$$\mathfrak{D}_0 \subset \mathfrak{D}\, , \quad \mathfrak{D} \subset \mathfrak{D}_0^*$$

and

$$\mathfrak{D}_0 \subset \mathfrak{D}_0^* , \quad \mathfrak{D} \subset \mathfrak{D}_0^* .$$

Actually, there are only two different ones among these spaces, as shown by the analogue of Theorem 32.2:

<u>Theorem 33.2</u>. $\mathfrak{D}^* = \mathfrak{D}_0, \quad \mathfrak{D}_0^* = \mathfrak{D}.$

To prove this theorem we proceed as in proving Theorem 32.2. Let ϕ be (1) in $\mathfrak{D}^*$, (2) in $\mathfrak{D}_0^*$. Then we should exhibit a sequence of functions $\phi^\nu(x)$ (1) in $\mathfrak{D}_0'$, (2) in $\mathfrak{D}'$ such that $\phi^\nu \to \phi$, $D\phi^\nu \to D\phi$.

We use "shifted" mollifiers by setting

$$\phi^\nu(x) = J_\pm^\nu \phi(x) = \int j^\nu(x-x' \mp \frac{1}{\nu})\phi(x')dx' .$$

Note that in the first case $j^\nu = 0$ when $x = 0$ and $x' \geq 0$ since then the argument $-x' - \frac{1}{\nu}$ of the kernel j^ν is outside its support; consequently $\phi^\nu(0) = 0$ so that $\phi^\nu(x)$ is in $\mathfrak{C}_0'$. In the second case we have $j^\nu = 0$ when $x' = 0$ and $x \geq 0$; consequently, the function j^ν considered as a function of x' belongs to $\mathfrak{D}_0'$. Keeping these facts in mind one may literally carry over the arguments of the proof of Theorem 32.2 to proving the present Theorem 33.2.

Thus we have to deal with only two operators, D in $\mathfrak{D}$ and $-D_0$ in $\mathfrak{D}_0$, which are strictly adjoint to each other. Using these operators we can form the two strictly self-adjoint operators $-D^2 = -DD_0$ and $-D_0^2 = -D_0D$ defined in appropriate dense spaces. The functions on which the first of these operators is applicable satisfy the first boundary condition $\phi(0) = 0$; those on which the second one is applicable satisfy the second boundary condition $D\phi(0) = 0$.

The spectral representations of these operators are given respectively by

$$\phi(x) = \int_0^\infty \sin \mu x \, \xi(\mu) d\mu ,$$

$$\xi(\mu) = \frac{2}{\pi} \int_0^\infty \sin \mu x \, \phi(x) dx$$

and

$$\phi(x) = \int_0^\infty \cos \mu x \, \xi(\mu) d\mu ,$$

$$\xi(\mu) = \frac{2}{\pi} \int_0^\infty \cos \mu x \, \phi(x) dx .$$

The eigenvalues of these operators are given by μ^2 in both cases.

The operator iD in $\mathfrak{D}_0$ is evidently formally self-adjoint but not strictly so since its domain $\mathfrak{D}_0$ differs from the domain $\mathfrak{D}$ of its strict adjoint iD. We may wonder whether or not the operator iD_0 in $\mathfrak{D}_0$ can be extended to a strictly self-adjoint operator. If that were possible the domain $\hat{\mathfrak{D}}$ of this extension would be contained in the space $\mathfrak{D}$. For certainly the relation $\mathfrak{D}_0 \subset \mathfrak{D}$ would imply $\hat{\mathfrak{D}}^* \subset \mathfrak{D}_0^*$; but $\mathfrak{D}^* = \hat{\mathfrak{D}}$ by assumption and $\mathfrak{D}_0^* = \mathfrak{D}$ by Theorem 33.2. Now, let ϕ_1 be a function in $\hat{\mathfrak{D}}$ not in $\mathfrak{D}_0$; then the corollary to Lemma 3 implies that every function ϕ in $\mathfrak{D}$ is in $\hat{\mathfrak{D}}$. That is to say, $\hat{\mathfrak{D}} = \mathfrak{D}$. There is hence no extension of $\mathfrak{D}_0$ contained in $\mathfrak{D}$ other than $\mathfrak{D}$ itself; but the operator iD in $\mathfrak{D}$ is evidently not self-adjoint.

We formulate this result, due to von Neumann, as

Theorem 33.3. The operator iD in $\mathfrak{D}_0$ is formally self-adjoint but cannot be extended to a strictly self-adjoint operator.

This result shows that the requirement of strict self-adjointness is essential; it is not simply a matter of mathematical completeness which could always be attained once the operator is Hermitean.

34. Partial Differential Operators

Our treatment of ordinary differential operators can be carried over to partial differential operators nearly literally. We consider functions ϕ of n variables $x = x_1,\ldots,x_n$ defined in the whole space $-\infty < x_1 < \infty,\ldots,-\infty < x_n < \infty$. We introduce the spaces $\mathfrak{C}$ and $\mathfrak{C}'$ of continuous and piecewise continuous functions, $\dot{\mathfrak{C}}'$ of piecewise continuous functions with bounded support, $\mathfrak{C}_1'$ of functions with finite support having a piecewise continuous first derivative. For the functions in $\mathring{\mathfrak{C}}'$ the integral

$$(\phi,\phi) = \int |\phi(x)|^2 dx$$

is defined where $dx = dx_1 \ldots dx_n$ and the integration is extended over the whole space. Using the same extension procedure that we have used for functions of a single variable, we may extend the space $\mathfrak{C}$ to a complete space of ideal functions, the Hilbert space $\mathfrak{H}$. Clearly, $\mathfrak{C}_1'$ is dense in $\mathfrak{H}$.

We introduce the space $\mathfrak{D}'$ of all functions in $\mathfrak{H}$ which have a piecewise continuous derivative which in turn is in $\mathfrak{H}$. We then define the operator D which transforms the function ϕ in $\mathfrak{D}'$ into the system of functions

$$D\phi(x) = \{\partial_1\phi(x),\ldots,\partial_n\phi(x)\} = \{\frac{\partial}{\partial x_1}\phi(x),\ldots,\frac{\partial}{\partial x_n}\phi(x)\}$$

which are piecewise continuous and in $\mathfrak{H}$. In the following it is convenient to consider a set of n functions as a single entity. We set

$$\psi = \{\psi_1,\ldots,\psi_n\}$$

and denote the spaces of functions $\psi(x)$ whose components are in $\mathfrak{C}$, $\mathfrak{D}$, $\mathfrak{H}$ by $\mathfrak{C}\cdot$, $\mathfrak{D}\cdot$, $\mathfrak{H}\cdot$; of course $\mathfrak{H}\cdot$ is a Hilbert space. The unit form in $\mathfrak{H}\cdot$ is

$$(\psi,\psi) = \int |\psi(x)|^2 dx \quad \text{where} \quad |\psi|^2 = \bar{\psi}\cdot\psi = \bar{\psi}_1\psi_1+\ldots+\bar{\psi}_n\psi_n \ .$$

In the space $\mathfrak{D}'\cdot$ we define the operator $D\cdot$ which transforms each function ψ in $\mathfrak{D}'\cdot$ into the function

$$D\cdot\psi = \partial_1\psi_1+\ldots+\partial_n\psi_n = \frac{\partial}{\partial x_1}\psi_1+\ldots+\frac{\partial}{\partial x_n}\psi_n,$$

which belongs to $\mathfrak{C}'$ and $\mathfrak{H}$.

The operators $-D$ in $\mathfrak{D}'$ and $D\cdot$ in $\mathring{\mathfrak{D}}'\cdot$ as well as the operators $-D$ in $\mathring{\mathfrak{D}}'$ and $D\cdot$ in $\mathfrak{D}'\cdot$ are formally adjoint to each other; for, the identity

$$-\int \bar{\phi} D\cdot\psi \, dx = \int D\bar{\phi}\cdot\psi \, dx$$

holds whenever ϕ is in $\mathfrak{D}'$ and ψ in $\mathring{\mathfrak{D}}'\cdot$ or ϕ is in $\mathring{\mathfrak{D}}'$ and ψ in $\mathfrak{D}'\cdot$. Note that no boundary terms appear in the integration by parts since one of the two functions vanishes identically outside of a finite region.

Since thus the operator D in $\mathfrak{D}'$ and $\mathring{\mathfrak{D}}'$ as well as $D\cdot$ in $\mathfrak{D}'\cdot$ and $\mathring{\mathfrak{D}}'\cdot$ have formal adjoints defined in dense domains, they admit closures. The domains of these closures will be denoted by $\mathfrak{D}$, $\mathring{\mathfrak{D}}$, $\mathfrak{D}\cdot$, $\mathring{\mathfrak{D}}\cdot$. Clearly we have

$$(\phi, D\cdot\psi) = -(D\phi,\psi)$$

for ϕ in $\mathfrak{D}$, ψ in $\mathring{\mathfrak{D}}\cdot$ and for ϕ in $\mathring{\mathfrak{D}}$, ψ in $\mathfrak{D}\cdot$.

Note that by this closure process we have extended the partial differential operator $D = \{\partial_1,\ldots,\partial_n\}$ to a class of functions $\phi(x)$ which do not all possess partial derivatives in the strict sense. It would be possible, though, to characterize the extended operator D in the terms of the Lebesgue theory. It does not seem possible, however, to characterize the extended divergence $D\cdot$ as producing the sum of the functions $\partial_1\psi_1,\ldots,\partial_n\psi_n$, each being obtained by applying an extended differential operator. The approach to extending the operators' gradient and divergence as described here avoids this difficulty.

In analogy to Theorem 32.1 we formulate

<u>Theorem</u> 34.1. $\mathring{\mathfrak{D}} = \mathring{\mathfrak{D}}$.

This statement is proved in nearly literally the same way as Theorem 32.1 was proved. We need not give details. The counterpart, $\mathfrak{D}\cdot = \mathring{\mathfrak{D}}\cdot$, of this statement is also true. We shall not attempt to prove this statement directly; its validity will eventually be ascertained without a special effort.

Since the operators $D\cdot$ in $\mathfrak{D}\cdot$ and $\breve{\mathfrak{D}}\cdot$ possess formal adjoints in dense domains they possess strict adjoints in domains $\mathfrak{D}*$ and $\mathring{\mathfrak{D}}*$. We have omitted the dot as multiplication symbol because these domains are spaces of single functions; in fact, these domains are extensions of $\mathfrak{D} = \mathring{\mathfrak{D}}$. Since $\mathring{\mathfrak{D}}\cdot \subset \mathfrak{D}'\cdot$ we have $\mathring{\mathfrak{D}}* \supset (\mathfrak{D}\cdot)*$. The adjoint operator $(D\cdot)*$ defined in $\mathring{\mathfrak{D}}*$ will be denoted by $-D$ since it is an extension of $-D$ defined in $\mathfrak{D}$. We now formulate

<u>Theorem</u> 34.2. $\mathring{\mathfrak{D}}* = \mathfrak{D}* = \mathfrak{D}$.

Because of $\mathring{\mathfrak{D}}* \supset \mathfrak{D}* \supset \mathfrak{D}$ this statement is implied by $\mathfrak{D}* = \mathfrak{D}$. To prove the latter statement we use mollifiers as for the proof of Theorem 32.2. We define the mollifiers J^ν as the integral operator with the kernel

$$j^\nu(x-x') = \nu^{-n} j(\nu(x_1-x_1'))\ldots j(\nu(x_n-x_n')) \ .$$

This n-dimensional mollifier has the same properties as the one-dimensional one. For ϕ in $\mathfrak{H}$

$$||J^\nu\phi|| \le ||\phi|| \ , \quad ||J^\nu\phi-\phi|| \to 0 \quad \text{as} \quad \nu \to 0$$

and $J^{\nu}\phi(x)$ is in $\mathfrak{C}_1'$. Furthermore, for each point x the function $\psi(x') = \{j^{\nu}(x-x'),0,\ldots,0\}$ is in $\mathring{\mathfrak{D}}'$ and

$$D\cdot\psi(x') = \partial_1' j^{\nu}(x-x') .$$

Hence, for ϕ in $\mathring{\mathfrak{D}}*$,

$$\int j^{\nu}(x-x')\partial_1'\phi(x')dx' = -\int \partial_1' j^{\nu}(x-x')\phi(x')dx$$

$$= \int \partial_1 j^{\nu}(x-x')\phi(x')dx' = \partial_1 \int j^{\nu}(x-x')\phi(x')dx' .$$

Since a corresponding relation holds for each component we have proved the identity

$$J^{\nu}D\phi(x) = DJ^{\nu}\phi(x)$$

for ϕ in $\mathring{\mathfrak{D}}*$. From $J^{\nu}D\phi \to D\phi$ and from the definition of $\mathfrak{D}$ as the domain of the closure of D in $\mathfrak{D}'$ we infer that ϕ is in this domain; i.e., ϕ is in $\mathfrak{D}$.

In the same manner one could prove the identity $\dot{\mathfrak{D}}\cdot* = \mathfrak{D}\cdot$ but it is not necessary to do so. For this identity follows immediately from the identity $\mathring{\mathfrak{D}} = \mathfrak{D}*$, using a theorem of von Neumann which states that $A^{**} = A$. The identity $\mathring{\mathfrak{D}} = \mathfrak{D}*$ in turn follows from $\mathring{\mathfrak{D}} = \mathfrak{D}$ and $\mathfrak{D} = \mathfrak{D}*$.

In the same manner von Neumann's theorem, applied to the identity $\mathring{\mathfrak{D}}* = \mathfrak{D}*$, yields the identity $\mathring{\mathfrak{D}}\cdot = \mathfrak{D}\cdot$. This is the counterpart of Theorem 34.1 which we have claimed to be valid.

Having proved the fact that the operators D in $\mathfrak{D}$ and $-D\cdot$ in $\mathfrak{D}\cdot$ are strictly adjoint to each other we can assert that the operators

$$-D\cdot D \quad \text{and} \quad -DD\cdot$$

are strictly self-adjoint. The first of these operators is the negative

Laplacean, and the second one is on occasion used in the theories of elasticity or electro-magnetism. It follows from the general theory that these operators admit spectral representations.

Of course the spectral representations of these operators can be given explicitly with the aid of the Fourier transformation. But the aim of our theory is not just to show that these particular operators admit of such a representation. Our primary aim is to show this to be the case for more general though related operators whose spectral representation cannot be given explicitly as simply as that of $-D \cdot D$.

35. Partial Differential Operators with Boundary Conditions

In Section 33 we imposed on our functions of one variable the condition that they should vanish at one point and found that the domain of the closure of the operator D acting on such functions is changed by imposing this condition. We shall find that the situation is quite different if we impose the condition of vanishing at a point on functions of more than one variable.

We denote by $\mathfrak{D}_0'$ the space of piecewise continuously differentiable functions $\phi(x)$ in $\mathfrak{H}$ which are (identically) zero in a neighborhood of the origin

$$\phi(x) = 0 \quad \text{for } |x| \leq \rho^2 .$$

(If we required only $\phi(0) = 0$ as in the case $n = 1$, we would obtain a somewhat weaker result.)

Since the operator D in $\mathfrak{D}_0'$ is formally adjoint to $D\cdot$ in the domain $\mathring{\mathfrak{D}}'$, which is dense in $\mathfrak{H}$, it follows that it possesses a closure: D in $\mathfrak{D}_0$. We now formulate

Theorem 35.1. $\mathfrak{D}_0 = \mathfrak{D}$ for $n > 1$.

In other words, the imposition of the condition $\phi(0) = 0$ did not make any difference on the closure of the operator D in $\mathfrak{D}'$.

We introduce the function $\zeta(\rho)$ defined by

$$\zeta_\rho(x) = \begin{cases} = \varepsilon_\rho \log \frac{\rho}{|x|} & \text{for } \rho^2 \leq |x| \leq \rho \\ = 0 & \text{for } \rho \leq |x| \\ = 1 & \text{for } |x| \leq \rho^2 , \end{cases}$$

$\varepsilon_\rho = 1/\log \rho^{-1}$. Then we have for any function $\phi(x)$ in $\mathfrak{D}'$

$$||D\zeta_\rho\phi|| \leq ||\zeta_\rho D\phi|| + || \, |D\zeta_\rho|\phi|| \leq ||D\phi \,||_\rho + ||D\zeta_\rho|| \; M_\rho\phi$$

with

$$||D\phi||_\rho^2 = \int_{|x|\leq\rho} |D\phi|^2 dx ,$$

$$M_\rho\phi = \max_{|x|\leq\rho} |\phi(x)| .$$

Now we have

$$||D\zeta_\rho||^2 \leq \frac{\Omega_n}{n-2} \rho^{n-2} \varepsilon_\rho^2 \quad \text{if} \quad n > 2$$

$$= \varepsilon_\rho \quad \text{if} \quad n = 2 ,$$

where Ω_n is the surface of the n-sphere. In any case

$$||D\zeta_\rho\phi|| \to 0 \quad \text{as} \quad \rho \to 0 .$$

Consequently, the function $\phi_\rho = (1-\zeta_\rho)\phi$ which is zero in the neighborhood of $x = 0$ approximates the function ϕ in such a way that

$$||D\phi_\rho - D\phi|| \to 0 , \quad ||\phi_\rho - \phi|| \to 0 \quad \text{as } \rho \to 0 .$$

Thus Theorem 35.1 is proved.

The situation is quite different if we impose a boundary

condition not just at a point but at an $(n-1)$-dimensional part of the boundary of a region $\mathscr{R}$ in the x-space. Let us take a rectangular cell

$$\mathscr{R} : 0 \leq x_\nu \leq a_\nu , \quad \nu = 1,\dots,n$$

as such a region $\mathscr{R}$ and denote by $\mathfrak{D}'_0$ the space of those functions with piecewise continuous first derivatives in $\mathscr{R}$ which vanish on the part

$$\mathscr{B}_0 : x_\nu = 0, \quad \nu = 1,\dots,n$$

of the boundary $\mathscr{B}$ of $\mathscr{R}$. By $\mathring{\mathfrak{D}}\cdot'$ we denote the space of those vectors $\psi(x) = \{\psi_1(x),\dots,\psi_n(x)\}$ which have a piecewise continuous derivative and vanish on the remaining part of $\mathscr{B}$

$$\mathscr{B}'_0 : x_\nu = a_\nu , \ \nu = 1,\dots,n .$$

Clearly the operators D in $\mathfrak{D}'_0$ and $-D\cdot$ in $\mathring{\mathfrak{D}}\cdot'$ are formally adjoint to each other. Hence these operators possess closures in domains $\mathfrak{D}_0$ and $\mathfrak{D}$.

The analogue of Theorem 33.1° is the statement that the space $\mathfrak{D}_0$ is not the same as the space $\mathfrak{D}$ defined without imposing a boundary condition. It can easily be proved by using the inequality

$$\Big[\int\cdots\int_{x_n=0} |\phi(x)|^2 dx_1\dots dx_{n-1}\Big]^{1/2}$$

$$\leq \Big[a\int_{x_n\leq a} |D\phi(x)|^2 dx\Big]^{1/2} + \Big[a^{-1}\int_{x_n\leq a} |\phi(x)|^2 dx\Big]^{1/2}$$

in place of the inequality for $|\phi(x)|$ used for the proof of Theorem 33.1°. We shall not give details.

We denote by $\mathfrak{D}'_0$ and $\mathring{\mathfrak{D}}\cdot'$ the domains of the closures of the operators D in $\mathfrak{D}_0'$ and D· in $\mathfrak{D}\cdot'$. As the analogue of Theorem 33.2 we then state that <u>the operator</u> D <u>in</u> $\mathfrak{D}_0$ <u>and</u> D· <u>in</u> $\mathfrak{D}\cdot$ <u>are strictly adjoint to each other</u>. Again this can be proved in the same way as Theorem 33.2 was proved, by using a mollifier kernel with shifted arguments:

$$j^{\nu}(x-x') = \nu^{-n} j(\nu(x_1-x_1') - 1) \ldots j(\nu(x_r-x_r') - 1) .$$

Considered as a function of x' this function vanishes on the part $\mathscr{B}^0$ of the boundary provided x is in the region $\mathscr{R}$. At the same time the function $J^{\nu}\phi(x)$ vanishes on the part $\mathscr{B}_0$ of $\mathscr{R}$. The arguments used in proving Theorem 32.2 then yield the statement.

We have chosen the rectangular cell as region $\mathscr{R}$ and imposed a boundary condition only on the part $\mathscr{B}_0$ of its boundary because then the arguments used for the one-dimensional case carry over with nearly no modification. Actually, the corresponding statements hold for any region with a sufficiently smooth boundary when the boundary condition is imposed on any $-(n-1)$-dimensional part of the boundary or on the whole boundary. To prove the analogue of Theorem 33.2 one may employ mollifier kernels with arguments which are shifted by the addition of a function of x rather than by just a constant. We shall not carry out the details here. In any case it follows that the operator $-D\cdot D$, the negative Laplacean, is strictly self-adjoint in an appropriate dense space of functions and hence possesses a spectral representation.

Various modifications of the Laplacean can be shown to be strictly self-adjoint. For example, we may consider the operator $-D\cdot D+V$ where the operator V consists in multiplication by a continuous function $v(x)$. Clearly, if the function $|v(x)|$ is bounded the operator $-D\cdot D+V$ is strictly self-adjoint in the same domain in which $-D\cdot D$ is.

Suppose $v(x)$ is not bounded but non-negative. Then we may

invoke Theorem 31.1 proved in Section 31 to ascertain that in an appropriate dense domain the operator $-D\cdot D+V$ is strictly self-adjoint. For clearly, the function $v(x)$ may be regarded as the least upper bound of bounded continuous functions.

If the function $v(x)$ is not bounded below the self-adjointness is not so easily established; it might not even be true without imposing additional conditions on the type of boundary condition. In some such cases, however, the problem can be handled by writing the operator in the form

$$-(D + g)\cdot(D + g)$$

where the vector $g(x) = g_1(x),\ldots,g_n(x)$ is a given real-valued function of x. If $g(x)$ is differentiable the operator in question when applied to twice-differentiable functions can be written in the form

$$(-D + g)\cdot(D + g) = -D\cdot D + V$$

with

$$V = -D\cdot g + g\cdot D + g\cdot g \ .$$

If the function $|g(x)|$ is bounded the domains of the closed operators $D + g$ and $-D\cdot + g\cdot$ are the same as those of D and $-D\cdot$. Moreover, the operators $D + g$ and $(-D + g)\cdot$ in these domains are strictly adjoint to each other. This statement is proved in the same way as Theorem 32.2, without noticeable modifications. Consequently, the operator $(-D + g)\cdot(D + g)$ is strictly self-adjoint in an appropriate dense domain; in general, this domain is not the same as that of $-D\cdot D$.

An interesting special case arises if we take

$$g = \alpha \frac{x}{|x|}$$

with any constant α. Although this function is discontinuous at $x =$

0 it is bounded; the statements made above are therefore valid. Thus the operator

$$L = \left(-D + \alpha \frac{x}{|x|}\right) \cdot \left(D + \alpha \frac{x}{|x|}\right)$$

is strictly self-adjoint in an appropriate dense domain. Also, of course, this operator is non-negative. When applied to continuously twice-differentiable functions which vanish at the origin this operator may be written in the form

$$L = -D \cdot D - \alpha \frac{n-1}{|x|} + \alpha^2 \ ;$$

it is thus recognized as Schrödinger's energy operator for the hydrogen atom (in case $n = 3$) except for the addition of the constant α^2. In this manner the strictly self-adjoint manner of the Schrödinger operator $L-\alpha^2$ is established.

Incidentally, the function

$$\phi(x) = e^{-\alpha|x|}$$

is an eigenfunction of L with the eigenvalue $\lambda = 0$ since it satisfies the equation

$$\left(D + \alpha \frac{x}{|x|}\right) \phi = 0 \ .$$

Since the operator L is non-negative the value $\lambda = 0$ is its lowest eigenvalue. Thus it is seen that the Schrödinger operator has no eigenvalue less than $-\alpha^2$.

36. Partial Differential Operators with Discrete Spectra

In our discussion of operators with discrete spectra in Chapter V we assumed the operator to be bounded. Since differential operators are not bounded it is necessary to modify the theory of Chapter V so as to cover differential operators. This is easy in case the differential operator is non-negative (except for an additive constant).

Let F be a strictly self-adjoint operator in a dense domain $\mathfrak{H}_F$ for which

$$F \geq C$$

with an appropriate constant C. Then we say that F is "of finite rank below λ" if there are vectors $Z^{(1)},\ldots,Z^{(g)}$ such that

$$\lambda||\Phi||^2 \leq \sum_{\gamma=1}^{g} |Z^\gamma,\Phi|^2 + (\Phi,F\Phi) \tag{36.1}$$

for all Φ in $\mathfrak{H}_F$. We maintain that <u>the spectrum of such an operator is discrete below</u> λ.

In proving this statement we may assume $F \geq 1$, $\lambda \geq 1$, without restriction. Let B be the inverse of F which exists by virtue of the operational calculus; then the square root $\sqrt{B}$ is defined and satisfies the relation $\sqrt{B}F\sqrt{B} = 1$, when applied to vectors Ψ for which $\sqrt{B}\Psi$ is in $\mathfrak{H}_F$. Setting $\Phi = \sqrt{B}$ in the inequality (36.1) above, we find the inequality

$$(\Psi,B\Psi) \leq \sum_{\gamma=1}^{g} |\Xi^\gamma,\Psi|^2 + \lambda^{-1}(\Psi,\Psi)$$

with $\Xi^\gamma = \sqrt{1/\lambda}\sqrt{B}\, Z^\gamma$ to be valid for all Ψ with $\sqrt{B}\Psi$ in $\mathfrak{H}_F$ and hence, by closure, for all Ψ in $\mathfrak{H}$.

From the theory of Chapter V we may now conclude that the spectrum of the operator B is discrete above λ^{-1}. Consequently, <u>the spectrum of the operator</u> F <u>is discrete below</u> λ.

We shall say the spectrum of F is "discrete" if the eigenspace of every finite interval is finite-dimensional. If the sequence of eigenvalues with multiplicity is infinite these eigenvalues tend to infinity. From the statement just proved we may then immediately conclude the validity of the

Corollary: Suppose inequality (0) holds for every value of λ (with appropriate vectors Z depending on λ). Then the spectrum of F is discrete.

We shall first consider a partial differential operator in a bounded domain and prove that its spectrum is discrete.

For simplicity we assume the operator to be simply the negative Laplacean, $-D\cdot D$, acting on functions $\phi(x)$ defined in a rectangular cell $0 \le x_\nu \le a_\nu$, $\nu = 1,\dots,n$, and satisfying the boundary condition $\phi = 0$ on $x_\nu = 0$, $\nu = 1,\dots,n$.

The following arguments, which go back to F. Rellich, can be carried over to more general operators, more general regions and more general boundary conditions provided that the coefficients of the differential operator are not singular and that the region is bounded and has a smooth boundary.

A major tool in the proof of discreteness is

Poincaré's inequality:

$$||\Phi||^2 \le R^{-1} |1,\Phi|^2 + \frac{d^2}{2} ||D\Phi||^2$$

where $R = a_1 \dots a_n$ is the volume of the cell $\mathscr{R}$ and $d = [a_1^2+\dots+a_n^2]^{1/2}$ its diagonal.

To prove this inequality we connect any two points x and x' in $\mathscr{R}$ by a zig-zag path, going from $x = x^{(1)} = (x_1,\dots,x_n)$ to $x^{(2)} = (x_1',x_2,\dots,x_n)$, then to $x^{(3)} = (x_1',x_2',x_3,\dots,x_n)$ and finally to $x^{(n+1)} = (x_1',\dots,x_n') = x'$. Evidently

$$|\phi(x^{(\nu+1)})-\phi(x^{(\nu)})| = \left|\int_{x^{(\nu)}}^{x^{(\nu+1)}} (\partial_\nu\phi(\tilde{x}))\,d\tilde{x}\right|$$

$$\le \left[a_\nu \int_{x^{(\nu)}}^{x^{(\nu+1)}} |\partial_\nu\phi|^2 dx_\nu\right]^{1/2} ;$$

hence

$$|\phi(x')-\phi(x)|^2 \leq d^2 \sum_{\nu=1}^{n} a_\nu^{-1} \int_{x^{(\nu)}}^{x^{(\nu+1)}} |\partial_\nu \phi|^\nu dx_\nu$$

$$\leq d^2 \sum_{\nu=1}^{n} a_\nu^{-1} \int_{x^{(\nu)}}^{x^{(\nu+1)}} |D\phi|^2 dx_\nu \ .$$

Integrating this inequality with respect to x and x' over $\mathscr{R}$ we obtain

$$2R||\Phi||^2 - 2|1,\Phi|^2 \leq d^2 R||D\Phi||^2$$

as claimed.

We apply this inequality to any of the sub-cells R_γ of R obtained by dividing each side into R equal parts. Letting $\eta_\gamma(x) = 1$ in $\mathscr{R}_\gamma$, $= 0$ outside $\mathscr{R}_\gamma$ we obtain

$$||\Phi||_\gamma^2 \leq R^{-1}k^n|\eta_\gamma,\Phi|^2 + \frac{d^2}{2k^2}||D\Phi||_\gamma^2$$

in obvious notation. Addition over all such cells gives the relation

$$||\Phi||^2 \leq R^{-1}k^n \sum_{\gamma=1}^{k^n} |\eta_\gamma,\Phi|^2 + \frac{d^2}{2k^2}||D\Phi|| \ ,$$

which is of the form (36.1) with $\lambda = 2k^2/d^2$ and $z^{(\gamma)} = \lambda R^{-1/2}k^{n/2}$, since

$$(\Phi,F\Phi) = -(\Phi,D\cdot D\Phi) = (D\Phi,D\Phi) \ .$$

Since k is arbitrary it is thus proved that <u>the operator $-D\cdot D$ is finite-dimensional below every value λ and hence that its spectrum is discrete</u>. Since the space $\mathfrak{H}$ is not finite-dimensional the sequence of eigenvalues is infinite; it follows that <u>these eigenvalues tend to infinity</u>.

Although the result thus obtained pertains only to finite regions, the argument that led to it can also be used for infinite regions. We shall prove the

Theorem. Suppose the function $v(x)$ tends to infinity as $|x|$ tends to infinity. Then the spectrum of the operator $-D\cdot D+V$ (acting on functions $\phi(x)$ defined in the whole x-space) is discrete; its eigenvalues tend to infinity.

To prove this statement we assume $v \geq 0$ throughout and let ω be an arbitrary positive number and let $\mathscr{R}$ be a region such that $v(x) \geq \omega$ when x is outside $\mathscr{R}$. To the inequality

$$(36.2) \qquad \omega||\Phi||_R^2 \leq \sum_{\gamma=1}^{g} |z^\gamma,\Phi|^2 + ||D\Phi||^2$$

obtained from Poincaré's inequality we add

$$\omega||\Phi||^2_{\mathscr{R}*} \leq (\Phi,V\Phi)_{\mathscr{R}*} \leq (\Phi,V\Phi)$$

where $\mathscr{R}*$ is the complement of $\mathscr{R}$. Thus we obtain

$$\omega||\Phi||^2 \leq \sum_{\gamma=1}^{g} |z^g,\Phi|^2 + (D\Phi,D\Phi) + (\Phi,V\Phi)$$

which is the desired inequality (36.1) since

$$(D\Phi,D\Phi) + (\Phi,V\Phi) = (\Phi,(-D\cdot D + V)\Phi) .$$

Thus the statement follows.

Finally, we shall consider a differential operator whose spectrum can be shown to be partially discrete on the basis of our criteria, namely the Schrödinger operator

$$F-\alpha^2 = (-D+g)\cdot(D+g) - \alpha^2 , \quad g(x) = \alpha x/|x| ,$$

acting on functions defined in the whole x-space.

Here we make use of the fact that the form $(\Phi,F\Phi)$ can be expressed in the form

$$(\Phi,F\Phi) = ((D+g)\Phi,(D+g)\Phi) \ .$$

We introduce a continuously differentiable function $\zeta(x)$ which vanishes for $|x| \leq \rho/2$ and equals 1 for $|x| \geq \rho$ and for which $0 \leq \zeta \leq 1$ throughout. Then we have

$$||(D+\zeta g)\Phi|| \leq \{||(D+g)\Phi|| + ||(1-\zeta)g\Phi||\}^2$$

$$\leq (1+\varepsilon)||(D+g)\Phi||^2 + (1+\varepsilon^{-1})\alpha^2||\Phi||^2_{R_\rho} \ .$$

Further,

$$||(D+\zeta g)\Phi||^2 = ||D\Phi||^2 - (\Phi,(D\cdot\zeta g)\Phi) + ||\zeta g\Phi||^2 \ ,$$

as verified by integration by parts. Now, to a given $\delta > 0$ we choose ρ so large that $D\cdot g = -(n-1)\alpha|x|^{-1} \geq -\alpha\delta/2$ and $|D\zeta| \leq |\alpha|\delta/2$, so that $D\cdot\zeta g \geq -\delta\alpha^2$. Then we obtain

$$(1+\varepsilon)||(D+g)\Phi||^2 \geq ||D\Phi||^2_{\mathscr{R}_\rho} - (2+\varepsilon^{-1})\alpha^2||\Phi||^2_{\mathscr{R}_\rho} + (1-\delta)\alpha^2||\Phi||^2.$$

Next we apply the inequality (36.2) to the region $|x| \leq 2\rho$ having chosen $\omega > (2+\varepsilon^{-1})\alpha^2$ so that we obtain

$$(1+\varepsilon)||(D+g)\Phi||^2 \geq (1-\delta)\alpha^2||\Phi||^2 - \sum_{\gamma=1} |z^\gamma,\Phi|^2 \ .$$

Since $(1+\varepsilon^{-1})(1-\delta)\alpha^2$ can be made arbitrarily close to α^2 we have found that <u>the spectrum of the operator</u> $F-\alpha^2$ <u>is discrete below any negative number</u>. This is a well-known property of the Schrödinger operator; but our derivation of this property evidently does not very specifically depend on the special form of the function $g(x)$; it would yield the same result for a wide class of such functions for which the spectral representation of the operator $(-D+g)\cdot(D+g)$ cannot be given explicitly.

CHAPTER VIII

PERTURBATION OF SPECTRA

37. Perturbation of Discrete Spectra

The method of perturbation is one of the most effective methods of determining approximately the spectral representation of a given operator. The method is applicable if the operator L in question is sufficiently near another operator L_0 whose spectral representation is known - or partly known. One then introduces an operator L_ε which depends on a parameter ε in such a way that it reduces to L for $\varepsilon = 1$ and to L_0 for $\varepsilon = 0$. Then one tries to expand the spectral representers of L_ε in powers of ε. The approximate representers of L_ε are obtained by breaking off the series after a few terms and then setting $\varepsilon = 1$.

If the undisturbed operator L_0 has a point eigenvalue λ_0, one may assume that the disturbed operator L_ε possesses also a point eigenvalue λ_ε which with an appropriately chosen eigenfunction Φ_ε possesses an expansion with respect to powers of ε. If this attempt succeeds, at least partial description of the spectral representation of the disturbed operator is attained. In the present section we shall be concerned with this problem of the perturbation of a point eigenvalue and closely related questions.

Let us first describe the perturbation procedure in a formal manner. We suppose that the operator L_ε is analytic in ε, and thus admits an expansion

$$L_\varepsilon = L_0 + \varepsilon L_1 + \ldots$$

(with bounded $L_1, L_2, \ldots$). Then we try to determine expansions

$$\lambda = \lambda_0 + \varepsilon\lambda_1 + \ldots, \quad \Phi_\varepsilon = \Phi_0 + \varepsilon\Phi_1 + \ldots$$

of an eigenvalue and an eigenvector of L_ε which reduce to a given eigenvalue λ_0 and a given eigenvector Φ_0 of L_0. Assuming the existence of such expansions we may write down a sequence of equations satisfied by the terms in it:

$$(L_0-\lambda_0)\Phi_0 = 0$$

$$(L_0-\lambda_0)\Phi_1 = -(L_1-\lambda_1)\Phi_0,$$

$$(L_0-\lambda_0)\Phi_2 = -(L_1-\lambda_1)\Phi_1 - (L_2-\lambda_2)\Phi_0 ,$$

$$\ldots\ldots$$

The question arises whether or not these equations allow us to determine the terms $\lambda_1, \lambda_2, \ldots$ and the vectors $\Phi_1, \Phi_2, \ldots$. If this is possible it remains to be shown that the series formed with these terms converge.

First we note that for every vector Φ

$$(\Phi_0, (L_0-\lambda_0)\Phi) = 0$$

holds if, as we assume, the operator L_0 is Hermitean. This fact implies that the right-hand sides of the sequence of equations above are orthogonal to Φ_0,

$$(\Phi_0, (L_1-\lambda_1)\Phi_0) = 0 , \quad (\Phi_0, (L_1-\lambda_1)\Phi_1) + (\Phi_0, (L_2-\lambda_2)\Phi_0) = 0 , \ldots .$$

Clearly, then, we can express successively the expansion coefficients λ_n of the eigenvalues in terms of the vectors $\Phi_0, \ldots, \Phi_{n-1}$. Suppose the right-hand side of the equation

$$(L_0-\lambda_0)\Phi_n = \Psi_n$$

has thus been made orthogonal to Φ_0; is it possible to determine Φ_n from this equation? This is not at all always possible, but it is possible if λ_0 is a <u>simple isolated</u> eigenvalue of L_0; i.e., if the eigenspace of L_0 associated with a sufficiently small open interval containing λ_0 consists just of the multiples of Φ_0. For, in this case, as is seen from the functional calculus, the restriction of $L_0-\lambda_0$ to the orthocomplement of Φ_0 has a bounded inverse. Therefore the above equation always has a solution, unique except for the addition of a multiple of Φ_0. This multiple can be made unique (for small enough ε at least) by adding the condition

$$(\Phi_0,\Phi_\varepsilon) = 1$$

together with

$$(\Phi_0,\Phi_0) = 1$$

so that

$$(\Phi_0,\Phi_n) = 0, \quad n = 1,2,\dots .$$

Because of its linearity this condition is preferable to the condition $||\Phi_\varepsilon|| = 1$, which of course could be achieved afterwards.

Instead of trying to prove the convergence of the resulting series for λ_ε and Φ_ε directly, one can proceed more effectively as follows: one allows the parameter ε to be complex, never mind that then the Hermitean character of L_ε is lost. Next one establishes the unique existence of solutions Φ_ε and λ_ε of the equations $(L_\varepsilon-\lambda_\varepsilon)\Phi_\varepsilon = 0$, $(\Phi_0,\Phi_\varepsilon) = 1$ for sufficiently small $|\varepsilon|$. Moreover, one establishes the unique existence of solutions Φ'_ε and λ'_ε of the equation

$$(L_\varepsilon-\lambda_\varepsilon)\Phi'_\varepsilon = -(L'_\varepsilon-\lambda'_\varepsilon)\Phi_\varepsilon \ , \ (\Phi_0,\Phi'_\varepsilon) = 0$$

with $L'_\varepsilon = dL_\varepsilon/d\varepsilon$, which would hold for the derivatives $\Phi'_\varepsilon = d\Phi_\varepsilon/d\varepsilon$, $\lambda'_\varepsilon = d\lambda_\varepsilon/d\varepsilon$ if they existed. Next one proves that the difference quotients of Φ_ε and λ_ε do approach these solutions Φ'_ε and λ'_ε, so that indeed Φ_ε and λ_ε have derivatives. Evidently, these derivatives are independent of the direction in which the difference quotients are taken. Hence Φ_ε and λ_ε are analytic in ε and, consequently, possess power series expansions. That these expansions are the same as those obtained before is clear from the uniqueness of their construction. Thus convergence of these series is established.

We shall carry out the main step in this argument not just for the problem of the disturbance of a single point eigenvalue as described. Rather, we shall carry out this step for a more general problem.

For simplicity we shall assume in the following that L_ε is of the form

$$L_\varepsilon = L_0 + \varepsilon V$$

where V is bounded (not necessarily Hermitean).

In the following we shall either omit the subscript ε or replace it by 1.

We assume that $\mathscr{I}_0$ is an isolated segment of the spectrum of L_0. By this we mean that the eigenspace of the operator L_0 associated with a slightly larger interval $\mathscr{I}$ is the same as that associated with $\mathscr{I}_0$, so that the spectrum is empty in the intervening intervals. Then we maintain that there exists an operator U^- which transforms the eigenspace $\mathfrak{C}_0$ of L_0 associated with $\mathscr{I}_0$ into an eigenspace $\mathfrak{C}$ of the operator L; moreover, we assume that U^- possesses an inverse U^+ which transforms $\mathfrak{C}$ back into $\mathfrak{C}_0$. This being so, the operator

$$\tilde{L} = U^+ L U^-$$

transforms the space $\mathfrak{C}_0$ into $\mathfrak{C}_0$, but in such a way that any eigenvector $\tilde{\Phi}_1$ of $\tilde{L}$ with the eigenvalue λ leads to an eigenvector

$$\Phi_1 = U^- \tilde{\Phi}_1$$

of L with the same eigenvalue λ. More generally, if, with reference to a spectral representation of the operator $\tilde{L}$ in $\mathfrak{C}_0$ the vector $U^+\Phi$ in $\mathfrak{C}_0$ has the spectral representer $\Psi(\lambda)$, the representation

$$\Phi \Leftrightarrow \Psi(\lambda)$$

affords a spectral representation of the vector Φ in the eigenspace $\mathfrak{C}$ of L. In other words, the problem of the spectral representation of the operator L in $\mathfrak{C}$ is reduced to that of the spectral representation of $\tilde{L}$ in $\mathfrak{C}_0$.

If in particular the space $\mathfrak{C}_0$ is finite-dimensional, the same is true of the eigenspace $\mathfrak{C}$ of L; the problem is reduced to that in a finite-dimensional space. If, for example, the spectrum of L_0 in $\mathfrak{C}_0$ consists of just one eigenvalue λ with the multiplicity r, the spectrum of L in $\mathfrak{C}$ consists of exactly r points counting each point with multiplicity. The problem of finding the r eigenvalues is thus reduced to a corresponding problem in a r-dimensional space.

It should be said that the problem of the splitting up of a multiple eigenvalue in a finite-dimensional space is far from trivial; it was first completely solved by Rellich in 1937. But the present approach shows that the problem of such a split-up in infinite-dimensional space can be reduced to a corresponding problem in finite-dimensional space.

How can one show the existence of such operators $U^{\pm}$ as described?

Of course, if the eigenspace $\mathfrak{C}_0$ is one-dimensional, consisting of the multiplesof an eigenvector Φ_0, the space $\mathfrak{C}$ is also one-dimensional consisting of the multiples of a vector Φ_1, which may be assumed to satisfy the relation $(\Phi_0, \Phi_1) = 0$. Then the operators $U^{\pm}$ given by

$$U^{\pm} = \Phi_0 \pm (\Phi_0, \Phi)(\Phi_0 - \Phi_1)$$

have the desired property, for then $U^{-}\Phi_0 = \Phi_1$ and $U^{+}\Phi_1 = \Phi_0$.

Thus the present problem covers the previous one.

Instead of the eigenspaces $\mathfrak{C}_0$ and $\mathfrak{C}$ we shall work with projectors P_0 and P which project into these spaces. We shall assume that P_0 is Hermitean, but we shall not make that assumption about P. Instead we shall require that P satisfy the two conditions

$$PP_0 = P \quad , \quad P_0P = P_0 \ . \tag{37.1}$$

Note that in the case the spaces $\mathfrak{C}_0$ and $\mathfrak{C}$ are one-dimensional, these conditions are satisfied if the projectors P_0, P are defined by

$$P_0 = (\Phi_0,\Phi)\Phi_0 \quad \text{and} \quad P\Phi = (\Phi_0,\Phi)\Phi_1 \ .$$

Also note that the two relations (37.1) imply the relation

$$P^2 = (PP_0)P = P(P_0P) = PP_0 = P \ ;$$

in other words, they imply that P is a projector.

The condition that $\mathfrak{C}$ is an eigenspace of L implies that $L\Phi$ lies in $\mathfrak{C}$, if Φ does, so that $LP\Phi$ always lies in $\mathfrak{C}$ and consequently $(1-P)LP\Phi = 0$. In other words, the relation

$$(1-P)LP = 0 \tag{37.2}$$

should hold. Conversely, the validity of this equation implies that P projects into a space $\mathfrak{C}$ transformed into itself by L. We insert $L = L_0+V$ in this equation and observe that

$$PL_0P = PP_0L_0P = PL_0P_0P = PL_0P_0 = PP_0L_0 = PL_0 \ ,$$

so that

$$(1-P)L_0P = L_0P - PL_0 = L_0(P-P_0) - (P-P_0)L_0 \ .$$

Accordingly, we may write equation $(1-P)LP = 0$ in the form

$$L_0(P-P_0) = (P-P_0)L_0 - (1-P)VP \ .$$

This equation, essentially due to C. Bloch (1958), will be solved to determine P.

We employ the function

$$\zeta(\lambda) \begin{cases} = \lambda^{-1} & \text{for } \lambda \text{ outside } \mathscr{I}_0 \\ = 0 & \text{for } \lambda \text{ in } \mathscr{I}_0 \end{cases}$$

and introduce the operator

$$\zeta(L_0) = Z_0 \, .$$

Since $P_0 = \eta(L_0)$ with

$$\eta(\lambda) \begin{cases} = 1 \text{ for } \lambda \text{ in } \mathscr{I}_0 \\ = 0 \text{ otherwise} \end{cases}$$

we clearly have

$$Z_0L_0 = L_0Z_0 = 1 - P_0 \ .$$

Multiplying our equation by Z_0 and observing

$$(1-P_0)(P-P_0) = P-P_0$$

we find

$$P-P_0 = Z_0(P-P_0)L_0 - Z_0(1-P_0-Q)V(P_0+Q) \, .$$

Suppose we have found an operator P which satisfies this equation together with $P^2 = P$, $PP_0 = P$, $P_0P = P_0$. Then $L_0(P-P_0) = (1-P_0)\{(P-P_0)L_0 - (1-P)VP\} = (P-P_0)L_0-(1-P)VP$ and hence $(1-P)LP = 0$.

Setting

$$f(Q) = Z_0 Q P_0 L_0 - Z_0(1-P_0-Q)V(P_0+Q)$$

we can write our equation in the form

$$Q = f(Q) .$$

It is assumed, without loss of generality, that $0 \in \mathcal{J}_0$. We denote by $\alpha\beta^{-1}$ and β^{-1} the maximum and minimum of $|\lambda|$ in $\mathcal{J}_0$ and $\bar{\mathcal{J}}$, respectively, where $\bar{\mathcal{J}}$ is the complement of $\mathcal{J}_0$ relative to the spectrum of L_0. It then follows that

$$||P_0 L_0|| \leq \alpha\beta^{-1} , \quad ||Z_0|| \leq \beta$$

and

$$||f(Q)|| \leq \alpha q + \beta(1+q)^2 ||V|| \quad \text{if } ||Q|| \leq q ,$$

$$||f(Q)-f(Q')|| \leq [\alpha + 2\beta(1+q)||V||]||Q-Q'||, \text{ if } ||Q||,||Q'|| \leq q .$$

We now require that the disturbing operator is sufficiently small:

$$||V|| < \frac{1-\alpha}{2\beta(1+q)} .$$

This bound is positive because $\mathcal{J}_0$ is an isolated segment of the spectrum. Then, with $\theta = \alpha q + 2\beta(1+q)||V|| < 1$,

$$||f(Q)|| \leq \alpha q + \frac{1-\alpha}{2}(1+q) \quad \text{and} \quad ||f(Q)-f(Q')|| \leq \theta||Q-Q'|| .$$

We now perform iterations

$$Q_{n+1} = f(Q_n)$$

beginning with $Q_0 = 0$, $Q_1 = -(1-P_0)VP_0$. To make sure that $||Q_1|| \leq q$ we require $q(1+q) \geq (1-\alpha)/2\beta$; to make sure that $||Q_n|| \leq q$ implies $||Q_{n+1}|| \leq q$ we require $q \geq 1$. Having done this we find $||Q_{n+1}-Q_n|| \leq \theta||Q_n-Q_{n-1}||$ and it is then clear that the sequence Q_n

converges in norm to a limit operator Q. Suppose the operator Q_n satisfies the condition $Q_n P_0 = Q_n$, $P_0 Q_n = 0$; then so does the operator Q_{n+1}. For then we have

$$f(Q_n)P_0 = Z_0 Q_n L_0 P_0 - Z_0(1-P_0-Q_n)V(P_0+Q_n)P_0 = f(Q_n)$$

$$P_0 f(Q_n) = 0 \quad \text{since} \quad P_0 Z_0 = 0 \ .$$

Hence the limit operator Q satisfies $QP_0 = Q$, $P_0 Q = 0$, i.e. the operator $P = P_0+Q$ satisfies $PP_0 = P$, $P_0 P = P_0$.

Relations $Q = QP_0$ and $P_0 Q = 0$ imply $Q^2 = QP_0 Q = 0$. The relation $Q^2 = 0$ is equivalent with $P^2 = P$ as verified from $PP_0 = PP_0 P = P_0$. Thus P is seen to be a projector. The space $\mathfrak{C}$ is defined as the range of P.

Relation $(1-P)LP = 0$, now established, which is equivalent to $LP = PLP$, shows that the operator L transforms $\mathfrak{C}$ into itself.

Finally, we set

$$U^{\pm} = 1 \mp Q \ .$$

Clearly, $U^- P_0 = P$, $U^+ P = P_0$, and $U^+ U^- = 1$. Hence U^- transforms $\mathfrak{C}_0$ onto $\mathfrak{C}$, while U^+ transforms $\mathfrak{C}$ onto $\mathfrak{C}_0$. Thus we have established the main step of the theory outlined.

We do not intend to carry out the details of the remaining steps which consist in showing that $P = P_\varepsilon$ has a unique derivative P'_ε with respect to ε. This can be done by arguments quite similar to those commonly employed to prove the differentiability of a solution of an ordinary differential equation with respect to a parameter. In many respects the existence of the solution that we have proved is more important than the possibility of expanding it in a power series.

We have seen that the Hermitean character of the perturbation V was not important; but it was important that this operator is bounded. The bound depends on the width of the gap $(1-\alpha)\beta^{-1}$ between the spectral interval $\mathcal{J}_0$ considered and the remaining part of the spectrum of L_0. If the minimal bound of V gets too large this remaining part of the spectrum may interfere with the part of the spectrum associated with the space $\mathfrak{C}$. There is a great variety in what the spectrum may suffer when this happens. If the remaining part of the spectrum is continuous, for example, it may happen that when the point eigenvalues coming from $\mathcal{J}_0$ reach the continuous spectrum they are absorbed and disappear.

38. Perturbation of Continuous Spectra

In the preceding section we have seen that the eigenspace belonging to a disconnected segment of the spectrum may move under perturbation, but keeps its dimension. If this dimension is infinite the nature of this segment of spectrum may change considerably; a point eigenvalue of infinite multiplicity may become a continuous spectrum; point eigenvalues may disappear when they come in contact with a continuous spectrum. In general one may say that the spectrum of an operator is very sensitive to disturbances. There are cases, however, where the spectrum does not change at all under certain perturbations. Such is the case, for example, if the undisturbed operator has a purely continuous spectrum and if the disturbing operator is sufficiently smooth in a sense to be explained.

We assume that the undisturbed operator L_0 has a continuous spectrum running from $-\infty$ to $+\infty$. (Later on, we shall show that the theory we shall develop automatically covers cases in which this spectrum covers only a finite or semi-infinite segment.) The vectors Φ of the Hilbert space may, therefore, be represented by functions $\psi(\lambda)$,

$$\Phi \Longleftrightarrow \psi(\lambda)$$

such that $L_0\Phi$ is represented by $\lambda\psi(\lambda)$:

$$L_0\Phi \Longleftrightarrow \lambda\psi(\lambda)$$

The "values" of the function ψ may be complex numbers; but ψ may also more generally be a complex-valued function of some accessory variables. In fact, we may simply say that the value of ψ may itself be a "vector" in an "accessory" Hilbert space. The expression

$$|\psi(\lambda)|^2 = \bar{\psi}(\lambda)\psi(\lambda)$$

is then understood as an inner product in this accessory space. Keeping this in mind, we assume for the unit form the expression

$$(\Phi,\Phi) = \int |\psi(\lambda)|^2 d\lambda;$$

here, and in the following, integration always extends from $-\infty$ to $+\infty$ unless otherwise stated.

The "smoothness" requirement we shall impose on the disturbing operator V will be expressed by saying that V should be represented by an integral operator,

$$V\Phi \Longleftrightarrow \int v(\lambda,\lambda')\psi(\lambda')d\lambda' ,$$

having a sufficiently smooth kernel. If the values of ψ are vectors in a Hilbert space, the values of V are bounded operators acting on vectors in this accessory space. The smoothness condition we shall impose on the kernel will be formulated later on. At present we discuss our aim in a formal manner and assume that <u>the disturbed operator</u>

$$L = L_0 + V$$

<u>has the same spectrum as</u> L_0. That is, we assume that the vectors Φ of the Hilbert space $\mathfrak{H}$ admit representations by square-integrable functions $\phi(\lambda)$,

$$\Phi \Longleftrightarrow \phi(\lambda) \;,$$

such that $L\Phi$ is represented by $\lambda\phi(\lambda)$,

$$L\Phi \Longleftrightarrow \lambda\phi(\lambda) \;.$$

The two representations will be referred to as the L_0- and the L-representation. Our aim then is to find a transformation from the L_0- into the L- representation and vice versa.

In cases in which the representing functions of two representations belong to the same class (the class of square-integrable functions in our case) it is convenient to effect the transformations with the aid of a pair of operators $U^{\pm}$ so chosen that the L- representer of Φ is the L_0- representer of U^+ , and the L_0- representer of Ψ is the L- representer of U^- . That is,

$$U^+\Phi \underset{0}{\Longleftrightarrow} \phi(\lambda) \qquad \Phi \Longleftrightarrow \phi(\lambda)$$

and

$$U^-\psi \Longleftrightarrow \psi(\lambda) \qquad \psi \underset{0}{\Longleftrightarrow} \psi(\lambda) \;.$$

From the first formula of the first line we may conclude

$$U^+L\Phi \underset{0}{\Longleftrightarrow} \lambda\phi(\lambda) \quad \text{since} \quad L\Phi \Longleftrightarrow \lambda\phi(\lambda) \;;$$

applying L_0 on this first formula, on the other hand, we obtain

$$L_0U^+\Phi \Longleftrightarrow \lambda\phi(\lambda) \;.$$

Thus,

$$U^+L = L_0U^+ \;. \tag{38.1}$$

Similarly, we derive

$$LU^- = U^-L_0 \tag{38.2}$$

from the second line. The equivalence of these two lines yields the formulas

$$U^+U^- = 1 \tag{38.3}$$

and

$$U^-U^+ = 1 \ , \tag{38.4}$$

which express the fact that U^+ is the inverse of U^- and vice versa.

Our aim now is to find a pair of operators $U^{\pm}$ which satisfy relations (38.1,2,3). Clearly, once such a pair has been found, the transformation of the L_0- into the L- representation (or vice versa) is established.

Setting $L = L_0+V$ we write equations (38.1), (38.2) in the form

$$L_0U^+ - U^+L_0 = U^+V \ ,$$

$$L_0U^- - U^-L_0 = -VU^- \ .$$

This form of the equation suggests that we should first investigate the solution of the equation

$$[L_0,Z] = L_0Z - ZL_0 = R$$

where R is a given operator. The theory of this equation is the basis for the treatment of the perturbation of continuous spectra.

Somewhat later on we shall describe a class of operators for which this equation has a solution. At present we shall assume that this is so. Evidently, this solution is not unique since any function of L_0 may be added to it, since any function of L_0 commutes with L_0. Later on we shall select a particular such solution which has particular desirable properties. We shall denote this solution, which depends linearly on R, by

$$Z = \Gamma R .$$

Our operators $U^{\pm}$ will be solutions of this basic equation with

$$R^{+} = U^{+}V, \quad R^{-} = -VU^{-} ;$$

but they will not be of the form $\Gamma R^{\pm}$ but rather of the form

$$U^{\pm} = 1 + \Gamma R^{\pm} .$$

This is natural since for $V = 0$ we will have $R^{\pm} = 0$ so that $U^{\pm} = 1$ as it should be. The operators $R^{\pm}$ are now to be found by solving the equations

$$R^{+} = V + (\Gamma R^{+})V , \quad -R^{-} = V + V(\Gamma R^{-}) .$$

We shall naturally try to solve these equations by iterations beginning with $R_0^{\pm} = 0$, $R_1^{\pm} = \pm V$. If we want to be sure that these iterations converge we must introduce a class (R) of operators R such that $R^{(1)}\Gamma R$ and $(\Gamma R^{(1)})R$ belong to this class if $R^{(1)}$ and R do; furthermore, this class should be complete with respect to a norm $||R||$ for which the inequalities

$$||R^{(1)}\Gamma R||, \; ||(\Gamma R^{(1)})R|| \leq ||R^{(1)}|| \; ||R||$$

hold. Then the iterations will converge provided $||V|| < 1$.

We proceed to discuss possible classes of operators R having the desired properties as described.

First of all we assume that the operators R are represented by integral operators with kernels $r(\lambda;\lambda')$:

$$R\Psi \underset{o}{\Longleftrightarrow} \int r(\lambda,\lambda')\psi(\lambda')d\lambda' \quad \text{when} \quad \Psi \underset{o}{\Longleftrightarrow} \psi(\lambda) ,$$

or simply

$$R \underset{o}{\Longleftrightarrow} r(\lambda;\lambda') .$$

Evidently, the operator $L_0Z - ZL_0$ is also an integral operator

$$[L_0,Z] = L_0Z - ZL_0 \Longleftrightarrow (\lambda-\lambda')z(\lambda;\lambda')$$

if Z is an integral operator

$$Z \Longleftrightarrow z(\lambda;\lambda') \ .$$

The relation $[L_0,Z] = R$ thus implies

$$(\lambda-\lambda')z(\lambda;\lambda') = r(\lambda;\lambda')$$

or

$$z(\lambda;\lambda') = (\lambda-\lambda')^{-1}r(\lambda;\lambda') \ .$$

Thus the kernel $z(\lambda;\lambda')$ would be singular unless $r(\lambda;\lambda) = 0$. In fact it is possible to introduce such singular integral operators and show that they have the desired properties provided the kernels $r(\lambda;\lambda')$ have an appropriate smoothness property. A smoothness property suitable for the purpose is Hölder continuity with respect to λ, λ', $1/\lambda$, and $1/\lambda'$. The details of justifying this statement involve considerable technicalities which we do not want to describe here. Let us denote by $\Gamma_0 R$ the operator Z with the kernel $(\lambda-\lambda')^{-1}r(\lambda;\lambda')$.

Actually, we shall not work with this solution of the equation $[L_0,Z] = R$; we shall rather work with the solution

$$\Gamma R$$

which transforms the vector Φ into the vector $\Gamma R\Phi$ represented by

$$\Gamma R\Phi \Longleftrightarrow \int (\lambda-\lambda')^{-1}r(\lambda;\lambda')\psi(\lambda')d\lambda' + i\pi r(\lambda,\lambda)\psi(\lambda) \ ,$$

and which may be regarded as an integral operator with the symbolic kernel

$$[(\lambda-\lambda')^{-1} + i\pi\delta(\lambda-\lambda')]r(\lambda;\lambda') \ .$$

The reasons why we shall work with this operator ΓR will become apparent later on.

We prefer to describe in detail a different class of operators which have the desired properties. The properties characterizing this class will be described in terms of their Fourier transforms

$$\rho(\sigma;\sigma') = \frac{1}{2\pi} \int\int e^{i\lambda\sigma - i\lambda'\sigma'} r(\lambda;\lambda')d\lambda d\lambda' .$$

The relationship between ρ and R will be indicated by

$$R \longleftrightarrow \rho(\sigma;\sigma') .$$

We now require that the functions ρ are absolutely integrable

$$\int\int |\rho(\sigma;\sigma')|d\sigma d\sigma' < \infty$$

and observe that this class $\mathscr{F}$ of functions is complete with respect to the norm

$$||\rho|| = \frac{1}{2\pi} \int\int |\rho(\sigma;\sigma')|d\sigma d\sigma'.$$

Let us first proceed in a formal manner, without asking under which conditions on the functions involved the operations are applicable.

If the kernel $z(\lambda;\lambda')$ corresponds to $\zeta(\sigma;\sigma')$, i.e. if

$$z(\lambda;\lambda') \longleftrightarrow \zeta(\sigma;\sigma') ,$$

we have, with $\partial_\sigma = \partial/\partial\sigma$,

$$(\lambda-\lambda')z(\lambda;\lambda') \longleftrightarrow -i(\partial_\sigma + \partial_{\sigma'})\zeta(\sigma;\sigma') .$$

Hence the equation

$$[L_0,Z] = R$$

goes over into the equation

$$-i(\partial_\sigma+\partial_{\sigma'})\zeta(\sigma;\sigma') = \rho(\sigma;\sigma') .$$

A solution of this equation is given by the kernel

$$\zeta(\sigma;\sigma') = \gamma\rho(\sigma;\sigma') = i\int_{-\infty}^{0} \rho(\sigma+\tau;\sigma'+\tau)d\tau;$$

for, $\partial_\sigma+\partial_{\sigma'}$ may be replaced by ∂_τ and then integration with respect to τ may be carried out. (It could be shown that the kernel $z(\lambda;\lambda')$ corresponding to this particular kernel $\gamma\rho(\sigma;\sigma')$ is exactly $[(\lambda-\lambda') + i\pi\sigma(\lambda-\lambda')]r(\lambda;\lambda')$ and thus improper.)

The kernel $\gamma\rho$ belongs to a class of kernels ζ for which the Hölmgren norm

$$||\zeta||_1 = \max \{\max_\sigma \int |\zeta(\sigma;\sigma')|d\sigma', \max_{\sigma'} \int |\zeta(\sigma;\sigma')|d\sigma\}$$

is finite, for evidently

$$||\gamma\rho||_1 \leq ||\rho|| .$$

Moreover, the kernel

$$\rho^{(2)}(\sigma;\sigma') = \int \rho^{(1)}(\sigma;\sigma'')\gamma\rho(\sigma'';\sigma')d\sigma'' ,$$

which corresponds to the operator $R^{(1)}\Gamma R$, belongs to the class $\mathscr{F}$ if ρ and ρ^1 do. For

$$||\rho^{(2)}|| \leq \int\int\int\int |\rho^{(1)}(\sigma;\sigma'')||\rho(\sigma''+\tau;\sigma'+\tau)|d\tau d\sigma'' d\sigma d\sigma'$$

$$= \int\int |\rho^{(1)}(\sigma;\sigma'')|d\sigma'' d\sigma \iint |\rho(\tilde{\sigma};\tilde{\sigma}')d\tilde{\sigma}d\tilde{\sigma}'$$

$$= ||\rho^{(1)}|| \; ||\rho|| .$$

Assigning the norm $||\rho||$ to the operator R:

$$||R|| = ||\rho|| ,$$

we may write the last result in the form

$$||R^{(1)}\Gamma R|| \leq ||R^{(1)}|| \; ||R|| .$$

In a similar way we derive

$$||(\Gamma R^{(1)})\, R|| \leq ||R^{(1)}|| \; ||R|| .$$

It is then clear that the iterations set up to solve the equations $R^{+} = V+(\Gamma R^{+})V$ and $R^{-} = -V-V(\Gamma R^{-})$ converge in the sense of the norm $||R||$ provided $||V|| < 1$.

From the fact that the kernel $\gamma\rho$ of the operators ΓR has a finite Hölmgren norm we deduce that the integral operator with this kernel transforms square-integrable functions of σ into square-integrable functions. Since the square integral is invariant under Fourier transformation, we may say that the operator ΓR is bounded, without referring to a kernel $z(\lambda;\lambda')$ of this operator. It thus follows that also the operators $U^{\pm} = 1 + \Gamma R^{\pm}$ are bounded.

Of the operator V we shall require that it be bounded in addition to being in class $\mathscr{F}$. Then the operators $R^{+} = U^{+}V$ and $R^{-} = VU^{-}$ are also bounded.

The kernel $\gamma\rho(\sigma;\sigma')$ need not be differentiable; still the relation $[L_0,\Gamma R] = R$ holds when applied to vectors which admit L_0. If the vector Φ is represented by the Fourier transform $\tilde{\phi}(\sigma)$ of its representer $\phi(\lambda)$, application of L_0 is represented by application of $i\partial_\sigma$. Suppose now $\tilde{\phi}(\sigma)$ admits this operator. Then, we maintain $\int\gamma\rho(\sigma;\sigma')\tilde{\phi}(\sigma')d\sigma'$ also admits $i\partial_\sigma$. To show this we need only show that $i\partial_\sigma$ is applicable weakly. We show a little more, namely that the relation

$$(L_0\Phi^{(1)},\Gamma R\Phi) - (\Phi^{(1)},(\Gamma R)L_0\Phi) = (\Phi^{(1)},R\Phi)$$

holds whenever Φ and $\Phi^{(1)}$ admit L_0 and R is bounded. To this end we need only approximate the kernel $\rho(\sigma;\sigma')$ in the $|| \ ||$-norm by a continuously differentiable one. For such a kernel the identity above can be proved in the same way as we had done it above in a formal way. The desired relation then results in the limit.

Thus we have proved: if the vector Φ admits L_0 the vectors $U^{\pm}\Phi$ admit L_0 and relations

$$L_0 U^+ \Phi = U^+(L_0+V)\Phi \ , \quad U^- L_0 \Phi = (L_0+V)U^- \Phi$$

hold.

Having constructed the operators $U^{\pm}$ we proceed to prove that they satisfy the relation $U^+U^- = 1$. In terms of the operators $R^{\pm}$ this relation takes the form

$$\Gamma(R^+ + R^-) + \Gamma R^+ \Gamma R^- = 0 \ .$$

To prove this relation we first write down the identity

$$R^+U^- = U^+VU^- = U^+R^-$$

or

$$R^+ + R^- + R^+\Gamma R^- + (\Gamma R^+)R^- = 0 \ .$$

Secondly, we make use of the identity

$$\Gamma(R^{(1)}\Gamma R^{(2)} + (\Gamma R^{(1)})R^{(2)}) = \Gamma R^{(1)}\Gamma R^{(2)}$$

which we shall prove presently. Combining it (for $R^{(1)} = R^+$, $R^{(2)} = R^-$) with the previous relation we indeed obtain the desired result $U^+U^- = 1$.

To prove the above identity we observe that in terms of transformed kernels it takes the form

$$- \int_{-\infty}^{0} \iint_{-\infty}^{0} \rho^{(1)}(\sigma+\tau';\sigma^{(2)})\rho^{(2)}(\sigma^{(2)}+\tau^{(2)};\sigma'+\tau'+\tau^{(2)})d\tau^{(2)}d\sigma^{(2)}d\tau'$$

$$- \int_{-\infty}^{0} \iint_{-\infty}^{0} \rho^{(1)}(\sigma+\tau''+\tau^{(1)};\sigma^{(1)}+\tau^{(1)})\rho^{(2)}(\sigma^{(1)};\sigma'+\tau'')d\sigma^{(1)}d\sigma^{(2)}d\tau''$$

$$= - \iint_{-\infty}^{0} \int_{-\infty}^{0} \rho^{(1)}(\sigma+\tau';\sigma''+\tau')\rho^{(2)}(\sigma''+\tau'';\sigma'+\tau'')d\tau''d\tau'd\tau'' ,$$

which is immediately verified by taking

$$\sigma'' = \sigma^{(2)}-\tau', \ \tau'' = \tau^{(2)}+\tau' \quad \text{and} \quad \sigma'' = \sigma^{(1)}-\tau'', \ \tau' = \tau^{(1)}+\tau''$$

as new variables in the integrals on the left side.

While it is thus seen that the identity $U^+U^- = 1$ is a consequence of the equations for $R^{\pm}$, identity $U^-U^+ = 1$ is not such a consequence unless additional requirements are imposed on V. The requirement

$$||V|| < \frac{1}{2}$$

may serve for this purpose. For this relation implies

$$||R^+|| < ||V|| \{1 + ||R^+||\} < \frac{1}{2} \{1 + ||R^+||\} ;$$

hence

$$||R^+|| < 1$$

and consequently

$$||\Gamma R^+||_1 < 1 ,$$

so that

$$U^+\Phi = 0 \quad \text{or} \quad \Phi = \Gamma R^+\Phi$$

implies $\Phi = 0$. Now

$$U^+(U^-U^+-1) = (U^+U^--1)U^+ = 0$$

so that

$$U^+(U^-U^+-1)\Phi = (U^+U^--1)U^+\Phi = 0$$

so that

$$(U^-U^+-1)\Phi = 0$$

follows for any Φ. Having thus established the remaining relation $U^-U^+ = 1$, we have attained our goal.

We just add that for practical purposes one will preferably use the expansions

$$R^+ = V + (\Gamma V)V + (\Gamma(\Gamma V))V + \ldots ,$$

$$R^- = -V + V\Gamma V - V\Gamma(V\Gamma V) + \ldots$$

to determine $R^{\pm}$ rather than iterations.

Two additional remarks should be made. First, we note that the theory developed covers the case in which the vectors Φ of the Hilbert space - now called $\mathfrak{H}_+$ - are represented by functions $\phi(\lambda)$ defined for $\lambda > 0$ only. The kernel $v(\lambda;\lambda')$ is then also defined only for $\lambda \geq 0$, $\lambda' \geq 0$.

We now define $\mathfrak{H}_+$ as the sub-space of $\mathfrak{H}$ whose vectors are represented by functions $\psi(\lambda)$ defined for all λ from $-\infty$ to $+\infty$ and for which

$$\psi(\lambda) = 0 \quad \text{for} \quad \lambda > 0 .$$

We extend the kernel $v(\lambda;\lambda')$ by setting

$$v(\lambda;\lambda') = 0 \quad \text{for} \quad \lambda < 0,\ \lambda' < 0 .$$

Then, clearly, the theory developed is applicable provided that <u>the</u>

kernel $v(\lambda;\lambda')$ for $\lambda \geq 0$, $\lambda' \geq 0$ is such that the extended kernel, defined for all λ,λ', is the Fourier transform of an absolutely integrable kernel $\nu(\lambda,\lambda')$.

We must make sure that the operators $R^{\pm}$ and $\Gamma R^{\pm}$ constructed in this theory also transform vectors of $\mathfrak{H}_+$ into $\mathfrak{H}_+$.

To this end we introduce the classes $\mathscr{F}_+$ of those kernels $\rho(\sigma;\sigma')$ which are Fourier transforms of kernels $r(\lambda;\lambda')$ which vanish for $\lambda \leq 0$ and for $\lambda' \leq 0$. Since

$$2\pi \max_{\lambda,\lambda'} |r(\lambda;\lambda')| \leq ||\rho|| \quad ,$$

it is clear that this class $\mathscr{F}_+$ is closed with respect to $||\rho||$.

We must further show that ΓR transforms $\mathfrak{H}_+$ into $\mathfrak{H}_+$ if R belongs to $\mathscr{F}_+$.

To this end we consider a vector Φ in the space $\mathfrak{H}$ whose representer, $\phi(\lambda)$, is of finite support and possesses a continuous second derivative. Then the Fourier transform $\tilde{\phi}(\sigma)$ of $\phi(\lambda)$ dies out at least like σ^{-2} as $|\sigma| \to \infty$. Using a vector Φ_- in $\mathfrak{H}_-$ with the same properties as Φ whose representer $\phi_-(\lambda)$ is zero in a neighborhood of $\lambda = 0$, we consider the inner product $(\Phi_-,\Gamma R\Phi)$ and show that it is zero. We have

$$\begin{aligned}(\Phi_-,\Gamma R\Phi) &= \iint \bar{\tilde{\phi}}_-(\sigma) \int_{-\infty}^{0} \rho(\sigma+\tau;\sigma'+\tau)\,d\tau\,\tilde{\phi}(\sigma')\,d\sigma' d\sigma \\ &= \int_{-\infty}^{0} \iint \bar{\tilde{\phi}}_-(\sigma-\tau)\,\rho(\sigma;\sigma')\,\tilde{\phi}(\sigma-\tau)\,d\sigma d\sigma' d\tau \quad ,\end{aligned}$$

where the interchange of the order of integration was permitted since $\tilde{\phi}(\sigma-\tau)$ and $\tilde{\phi}_-(\sigma-\tau)$ decay at least like τ^{-2} as $\tau \to -\infty$. Now, the Fourier transform $e^{-i\tau\lambda}\phi(\lambda)$ of $\tilde{\phi}(\sigma-\tau)$ is the representer of a vector $\Phi(\tau)$ in $\mathfrak{H}$ while the transform $e^{-i\tau\lambda}\phi_-(\lambda)$ of $\tilde{\phi}_-(\sigma-\tau)$ represents a vector $\Phi(\tau)$ in $\mathfrak{H}_-$. Consequently

$$\iint \bar{\tilde{\phi}}_-(\sigma-\tau)\rho(\sigma;\sigma')\tilde{\phi}(\sigma-\tau)d\sigma d\sigma' = (\Phi_-(\tau), R\Phi(\tau)) = 0$$

since $R\Phi(\tau) \perp \mathfrak{H}_-$. Thus, our statement $(\Phi_-, \Gamma R\Phi) = 0$ is proved. Since the spaces of the vectors Φ, Φ_- of the kind considered are dense in $\mathfrak{H}$ and $\mathfrak{H}_-$, it follows that $\Gamma R\Phi \perp \mathfrak{H}_-$ for all Φ in $\mathfrak{H}$; i.e., ΓR transforms $\mathfrak{H}$ into $\mathfrak{H}_+$.

This result implies that all the operators which are formed in the process of iterations as approximations to $R^{\pm}$ belong to $\mathcal{F}_+$. We conclude that the operators $R^{\pm}$ themselves belong to $\mathcal{F}_+$. Hence $R^{\pm}$ and $\Gamma R^{\pm}$ transform vectors of $\mathfrak{H}_+$ into $\mathfrak{H}_+$.

We should like to elaborate on the condition, imposed above, that the extended kernel $v(\lambda;\lambda')$ be the Fourier transform of an absolutely integrable kernel $-\nu(\sigma;\sigma')$. Since the Fourier transform of an absolutely integrable function is continuous, this condition implies that <u>the original kernel</u> $v(\lambda;\lambda')$ <u>is continuous and vanishes along the boundary</u> $\lambda = 0$ <u>and</u> $\lambda' = 0$ of its domain of definition. This continuity condition is quite significant. Although it is not necessary for the existence of transformations $U^{\pm}$ (of a wider class than here considered), some such condition is necessary.

To show this we shall consider <u>an example</u> of a kernel which violates our condition, namely the kernel

$$v(\lambda;\lambda') = \varepsilon b(\lambda)b(\lambda')$$

where the function $b(\lambda)$, defined for $\lambda \geq 0$, is absolutely integrable and bounded, but does not vanish for $\lambda = 0$. Moreover, we assume

$$\int_0^\infty |b(\lambda)|^2 d\lambda = 1 .$$

We maintain that for $\varepsilon > 0$ the operator $L = L_0+V$ possesses a point eigenvalue, so that clearly its spectrum is not simply the same as that of L_0, namely the semi-axis $\lambda \geq 0$. Such a point eigenvalue is easily exhibited.

Suppose there were such a point eigenvalue λ_ε with the eigenvector Φ_ε; then we would show

$$(L_0 - \lambda_\varepsilon)\Phi_\varepsilon = -V\Phi_\varepsilon$$

or

$$(\lambda - \lambda_\varepsilon)\phi_\varepsilon(\lambda) = \varepsilon c b(\lambda)$$

with

$$c = \int_0^\infty \overline{b(\lambda')}\, \phi_\varepsilon(\lambda) d\lambda' \ .$$

Assuming $\lambda < 0$, we would have

$$\phi_\varepsilon(\lambda) = c\varepsilon \frac{b(\lambda)}{\lambda - \lambda_\varepsilon}$$

and hence

$$c = c\varepsilon \int_0^\infty \frac{|b(\lambda')|^2}{\lambda' - \lambda_\varepsilon} d\lambda' \ .$$

Then $\lambda = \lambda_\varepsilon$ satisfies the equation

$$F(\lambda) = \int_0^\infty \frac{|b(\lambda')|^2}{\lambda' - \lambda} d\lambda' = \frac{1}{\varepsilon} \ .$$

A simple discussion, using $b(0) \neq 0$, shows that for $\lambda < 0$ the function $F(\lambda)$ on the left-hand side is positive and grows from 0 to ∞ as λ varies from $-\infty$ to 0. Obviously, there is exactly one solution $\lambda = \lambda_\varepsilon$ if $\varepsilon > 0$.

Suppose the function $b(\lambda)$ vanishes at $\lambda = 0$ after all, like a power of λ, say. Then our theory is applicable. Indeed, the function $F(\lambda)$ remains finite and the equation $F(\lambda) = 1/\varepsilon$ has no solution if ε is sufficiently small. If ε is sufficiently large there is again a solution. This does not contradict our theory since we had restricted the norm $||V||$ of V to be sufficiently small, but it shows that some such restriction is essential.

39. Scattering

One of the important features of the theory of perturbation of continuous spectra is its role in the description of scattering. The process of scattering is described with the aid of the operator

$$e^{-itL}$$

(where L may stand for the Hamiltonian energy operator of quantum theory). The aim is to relate the limit that this operator approaches as t (the time) tends to $-\infty$ with the limit approached as t tends to $+\infty$.

Note that the operator e^{-itL} gives the solution

$$\Phi(t) = e^{-itL}\Phi(0)$$

of the Schrödinger equation

$$i \frac{d}{dt} \Phi(t) = L\Phi(t) .$$

Actually, the operator e^{itL} does not approach a limit as $t \to \pm\infty$, but the operator

$$e^{itL_0}e^{-itL}$$

does. If a spectral representation of L_0 is employed, application of e^{+itL_0} is represented by multiplication by the "phase factor" $e^{+it\lambda}$, which for the present problem is insignificant.

The operator $e^{+itL_0}e^{-itL}$ can easily be described with the aid of the operators $U^{\pm}$. Since

$$L = U^{-}L_0U^{+}$$

we have

$$f(L) = U^{-} f(L_0) U^{+}$$

and hence

$$e^{itL_0} e^{-itL} = e^{itL_0} U^{-} e^{-itL_0} U^{+} = S(t) U^{+}$$

with

$$S(t) = e^{itL_0} U^{-} e^{-itL_0} = 1 + \Gamma_t R^{-}$$

where

$$\Gamma_t R^{-} = e^{itL_0} \Gamma R^{-} e^{-itL_0} .$$

It is easy to give the Fourier transform of the kernel of the operator $\Gamma_t R^{-}$. Since the Fourier transform of the kernel of any operator R is

$$\frac{1}{2\pi} \iint e^{i\lambda\sigma - i\lambda'\sigma'} r(\lambda;\lambda') \, d\lambda d\lambda' = \rho(\sigma;\sigma')$$

the transform of the kernel of $e^{itL_0} R e^{-itL}$ is

$$\frac{1}{2\pi} \iint e^{i(\lambda\sigma - \lambda'\sigma')} e^{it\lambda} r(\lambda;\lambda') e^{-it\lambda'} \, d\lambda d\lambda' = \rho(\sigma+t;\sigma'+t) .$$

The transform of $\Gamma_t R$ is therefore

$$\gamma_t \rho(\sigma;\sigma') = \int_{-\infty}^{0} \rho(\sigma+t+\tau;\sigma'+t+\tau) \, d\tau = \int_{-\infty}^{t} \rho(\sigma+\tau;\sigma'+\tau) \, d\tau .$$

Now this kernel tends to zero as $t \to -\infty$ and, as $t \to \infty$, to

$$\gamma_\infty \rho(\sigma;\sigma') = \int_{-\infty}^{\infty} \rho(\sigma+\tau;\sigma'+\tau) \, d\tau .$$

This is true even with respect to the Hőlmgren norm:

$$||\gamma_t\rho||_1 \to 0 \quad \text{as} \quad t \to -\infty$$

$$||\gamma_t\rho-\gamma_\infty\rho||_1 \to 0 \quad \text{as } t \to \infty ,$$

as is easily verified (by first verifying this for kernels of finite support). Evidently,

$$\gamma_\infty\rho(\sigma;\sigma') = \mu(\sigma-\sigma')$$

with

$$\mu(\sigma) = \int_{-\infty}^{\infty} \rho(\sigma+\tau;\tau)d\tau .$$

Consequently, application of the operator $\Gamma_\infty R$ corresponding to the kernel $\gamma_\infty\rho(\sigma;\sigma')$ on a vector Φ represented by $\phi(\lambda)$ consists in multiplication of $\phi(\lambda)$ by the factor

$$m(\lambda) = \int_{-\infty}^{\infty} e^{-i\lambda\sigma}\mu(\sigma)d\sigma.$$

The operator $S(t) = 1+\Gamma_t R^-$, in particular, approaches the operator

$$S_\infty = 1 + \Gamma_\infty R^-$$

which consists in multiplying just by the factor

$$\gamma_\infty(\lambda) = 1 + m(\lambda)$$

for the L_0- representation. This operator S_∞ is the "scattering operator". It transforms the limit U^+ of $e^{itL_0}e^{-itL}$ as $t \to -\infty$ into the limit $S_\infty U^+$ of this operator as $t \to +\infty$.

We recall that at the beginning of Section 38 we said that the values of the representing functions $\phi(\lambda)$ may themselves be functions of some accessory variables and that then the values of the kernels $v(\lambda;\lambda')$ or $r(\lambda;\lambda')$ are matrices or integral operators acting on functions of these accessory variables. In this sense, then, the scattering operator is also a matrix or an intergral operator acting on functions of these accessory variables. Aside from this, however, the scattering operator is a function of the operator L_0 and hence commutes with L_0.

Frequently, in place of the operator S_∞ the operator

$$U^- S_\infty U^+$$

is designated as the scattering operator. It evidently commutes with the disturbed operator L in place of L_0.

REFERENCES

Akhieser, N.I., and Glasmann, I.M. - Theorie der Linearen Operatoren im Hilbert-Raum, Akademie Verlag Berlin, 1954.

Dunford, N., and Schwartz, J.T. - Linear Operators, Interscience Publishers, Inc., New York, 1958.

Riesz, F., and Nagy, B.S. - Functional Analysis (translation), F. Ungar Publishing Company, New York, 1955.

Stone, M.H. - Linear Transformations in Hilbert Space, American Mathematical Society Colloquium Publications, New York, 1932.

Taylor, A.E. - Introduction to Functional Analysis, J. Wiley & Sons, New York, 1958.

INDEX

Graduate Texts in Mathematics

Soft and hard cover editions are available for each volume.

For information

A student approaching mathematical research is often discouraged by the sheer volume of the literature and the long history of the subject, even when the actual problems are readily understandable. The new series, Graduate Texts in Mathematics, is intended to bridge the gap between passive study and creative understanding; it offers introductions on a suitably advanced level to areas of current research. These introductions are neither complete surveys, nor brief accounts of the latest results only. They are textbooks carefully designed as teaching aids; the purpose of the authors is, in every case, to highlight the characteristic features of the theory.

Graduate Texts in Mathematics can serve as the basis for advanced courses. They can be either the main or subsidiary sources for seminars, and they can be used for private study. Their guiding principle is to convince the student that mathematics is a living science.

Vol. 1 TAKEUTI/ZARING: Introduction to Axiomatic Set Theory. vii, 250 pages. 1971.

Vol. 2 OXTOBY: Measure and Category. viii, 95 pages. 1971.

Vol. 3 SCHAEFER: Topological Vector Spaces. xi, 294 pages. 1971.

Vol. 4 HILTON/STAMMBACH: A Course in Homological Algebra. ix, 338 pages. 1971.

Vol. 5 MAC LANE: Categories for the Working Mathematician. ix, 262 pages. 1972.

Vol. 6 HUGHES/PIPER: Projective Planes. xii, 291 pages. 1973.

Vol. 7 SERRE: A Course in Arithmetic. x, 115 pages. 1973.

Vol. 8 TAKEUTI/ZARING: Axiomatic Set Theory. viii, 238 pages. 1973.

Vol. 9 HUMPHREYS: Introduction to Lie Algebras and Representation Theory. xiv, 169 pages. 1972.

Vol. 10 COHEN: A Course in Simple-Homotopy Theory. xii, 114 pages. 1973.

Vol. 11 CONWAY: Functions of One Complex Variable. xiv, 314 pages. 1973.

In preparation

Vol. 12 BEALS: Advanced Mathematical Analysis. xii, 236 pages approximately. Tentative publication date: September, 1973.

Vol. 13 ANDERSON/FULLER: Rings and Categories of Modules. xiv, 370 pages approximately. Tentative publication date: October, 1973.

Vol. 14 GOLUBITSKY: Stable Mappings and Their Regularities. xii, 212 pages approximately. Tentative publication date: October, 1973.

ISBN 0-387-9007
ISBN 3-540-9007